Ten-Minute Field Trips

Other books by Helen Ross Russell

City Critters
Clarion the Killdeer
Small Worlds: A Field Trip Guide
Soil: A Field Trip Guide
The True Book of Buds
The True Book of Springtime Tree Seeds
Winter: A Field Trip Guide
Winter Search Party

Ten-Minute Field Trips

A Teacher's Guide to
Using the School Grounds for Environmental Studies
Second Edition

BY HELEN ROSS RUSSELL
Science Consultant, Manhattan Country School

Illustrated by Klaus Winckelmann

National Science Teachers Association

National Science Teachers Association
Produced by Special Publications
1742 Connecticut Avenue, NW
Washington, DC 20009

Originally published by
J. G. Ferguson Publishing Company in 1973

ISBN 0-87355-098-6
PB-20
Library of Congress Catalog Card Number: 90-63468
Printed in the U.S.A.

TO THELMA J. GREENWOOD
a teacher with imagination who taught me much by example

About the Author ~~~~~~~~~~~~~~~~~~~~~~~~

HELEN ROSS RUSSELL started teaching young people when she was 19 and has never stopped. She has taught kindergarten through graduate school, in urban, small town, and suburban areas. For ten years, she was academic dean of the State College in Fitchburg, MA. She continues to actively teach at the Manhattan Country School and conducts teacher workshops from Maine to Arizona. Her love of the outdoors, her enjoyment of children, and her enthusiasm for life in all its forms make *Ten-Minute Field Trips* an exciting adventure in awareness. The author of numerous other books on the natural world, Dr. Russell has a uniquely open, inquisitive, and happy approach to learning, and enjoys a national reputation as an environmental educator of young children in an urban setting. Whether on a concrete school ground (indeed hundreds of trips in the book are directed to just such an area) or on a grassed and landscaped ground, *Ten-Minute Field Trips* gives a wealth of ideas for instruction in the best classroom of all—the world around us.

Cover Photo: A fire escape is an excellent resource for studying light, sound, magnets, and simple machines. Photo by Ray Pfortner, Manhattan Country School

Contents

Contents (*continued*)

Foreword ᵔᵔᵔᵔᵔᵔᵔᵔᵔᵔᵔᵔᵔᵔᵔᵔᵔᵔᵔᵔᵔᵔ

Education about the environment is key to the survival of our planet. Our nation has serious environmental problems and we need to collect the energy and creativity of young minds working on them. A widespread understanding of how natural systems work would greatly accelerate the process of enlightening and educating our youth and their families. How we see and explain the world and convey that knowledge to future generations determines the organic growth and development of our society.

In preparing our youth for the environmental challenges of today and the twenty-first century *Ten–Minute Field Trips* by Dr. Helen Ross Russell is a key resource for your collection. The organization of this book will guide you through a sequence of investigative experiences which are engaging without requiring elaborate equipment or vast knowledge about natural history. In designing this book Dr. Russell has demonstrated her extraordinary ability to present complex ideas and scientific information in a "user–friendly" manner.

My introduction to Dr. Russell's teaching philosophy and style was during my graduate studies at the University of Michigan in the early seventies. *Ten–Minute Field Trips* was required reading and study by all students committed to teach about the environment. I was intrigued by the notion of a school site and its surrounding neighborhood as an environmental studies laboratory where teachers and students actively investigate their natural and built environs.

I have used *Ten–Minute Field Trips* extensively as a primer in my work in New York public schools with elementary and junior high school students and teachers in the neighborhoods of Harlem and the South Bronx. The lessons contained in this book provided me with the framework to help urban young people see their inner city world with positive eyes and value the natural environs within it.

I was fortunate to witness Dr. Helen Ross Russell in action at the American Museum of Natural History in New York City. There in the Eastern Woodland Indians exhibit she guided a group of elementary school teachers through a concise, inspiring experience. Dr. Russell shared her style and teaching philosophy, masterfully demonstrating how to motivate young people to invest their time and creative energy in exploring how nature works. Each teacher walked away inspired with confidence, factual knowledge, and instructional techniques and skills that would empower them to become better teachers. I will always be thankful for having participated in that session and for the numerous collaborations that have followed.

The goal of environmental education is to nurture a citizenry who are environmentally astute and literate about the biophysical phenomena around them and the built environment. This educational process should occur through teaching, mass media, citizen action, or through a blend of strategies. We have the ability to alter our environment in positive or negative ways. The degree of impact we have on the environs of the world is greatly influenced by our level of knowledge, skills, attitudes, values and by our social and political systems.

Ten–Minute Field Trips is a crystallization of Dr. Russell's knowledge and teaching techniques and it invites all of us to participate in saving the planet for future generations. This volume will be a welcome resource aid to the formal and nonformal educational community attempting to meet their goal. Dr. Russell has touched the hearts and minds of thousands of teachers and students during her outstanding career and this volume is a tribute to her creative vision.

Talbert B. Spence
Director
Educational Programs
The New York Academy of Science

Preface

While this book emphasizes the use of the school grounds in environmental studies, such studies are not limited to that particular spot. The direction of the material is much more a matter of philosophy than of geography. Short close-to-home field trips are essential for understanding the environment. Hands-on activities in the local environment open new dimensions that cannot be found anywhere else. They also make classroom learning more meaningful and fun.

Sharing this excitement has always been one of my goals as an author. In fact, meeting the people who have read and used my books is one of the special rewards of being an author. Two years after *Ten–Minute Field Trips* was originally published, I was approached by a young woman whom I had never met. She wanted to express her appreciation of the book, which she told me she had used steadily since she purchased it as a text for a Bank Street College of Education course teaching social studies for early childhood.

I still glow when I think of this. Here obviously was a chain from my experiences with hundreds of classroom teachers and thousands of children, to a book, to a college professor with vision, to other classroom teachers whose children would know the joy of discovery.

Not only can short field trips enhance school curricula in a variety of subjects, many of the activities in *Ten–Minute Field Trips* lend themselves to day camp, resident camp, and nature center programs.

In addition, many parents may find this book useful. The introductory information on each topic is written for the non-scientist. It will help parents cope with a child scientist's never-ending questions while it makes suggestions for activities that parents can share with their children in their own community.

—Helen Ross Russell

Ten-Minute Field Trips was produced by NSTA Special Publications, Shirley Watt Ireton, managing editor; Cheryle Shaffer, Andrew Saindon, Christine Pearce, assistant editors; and Gregg Sekscienski, editorial assistant. Cheryle Shaffer was NSTA editor for *Ten-Minute Field Trips*. Phyllis Marcuccio is NSTA's assistant executive director for publications. Typesetting and printing at Automated Graphic Systems was handled by Susan Zimmerman. Russ Peltier of Association Graphics designed the cover.

Ten-Minute Field Trips

These nine year olds are going on a fact-finding field trip. When all students know the purpose of the trip and can experiment, measure, and record on their own, total participation and individual learning will follow.

Using the School Grounds for Environmental Studies

Regardless of whether we look on the environment in its narrowest sense of the immediate surroundings, or in its all-inclusive sense of this ecosystem—this spaceship, Earth—the area surrounding the school is by far the best starting point.

This means that the best possible facilities for teaching environmental studies are available to all schools, regardless of whether they are urban, suburban, or rural. By using school grounds effectively, teachers and children can learn about all the natural forces as they relate to their situation. Temperature changes, precipitation, air currents, pollution, the forces of disintegration and decomposition, plant and animal relationships, and people relationships are things that occur everywhere, but their intensity and effect vary with the locality.

School grounds are almost always reflectors of the neighborhood in which they are located. The hard-topped play area of a metropolitan school is touched by rain, wind, pollution, noise, and overcrowding in the same way as the streets that surround it; similarly, the bare soil of a hillside school develops gulleys in the same way as the farms that surround it.

From the teaching standpoint, school grounds have the added advantage of being easily accessible. Repeated trips to observe changes may be made in a day, or over a period of several days, or even a year. Not only does the proximity of the school grounds make meaningful observation possible, it provides the opportunity to initiate projects to improve the environment.

Thus, the school grounds can provide an opportunity to teach the three things needed if we are to develop responsible environmental action. These are:

1. an awareness and understanding of the interrelationships in the natural world;

2. a concern about the misuse of this planet;

3. a willingness and ability to initiate and support positive action on the basis of this knowledge.

In addition to these primary advantages, there are many other important advantages to using the school grounds as the main base of operation for field trips. For example, there is no scheduling problem; no waiting for a bus date; no need to hurry up a topic or to try to rekindle interest in a topic completed weeks before. Dangers are also minimized. There are no streets to cross, no transportation problems, and the same insurance that covers children in the school building covers them on the grounds.

The trip can be a natural outgrowth of the topic at hand: "Let's go outside and look at the rock outcropping at the corner of our school grounds." "When the sun comes out, maybe we could set up a shadow study." "Where do most of the earthworms on our school grounds live? How can we find out?" "What kinds of pollution problems affect us here at the corner of 81st Street and Broadway?"

Responses to challenges of this type are most often enthusiastic because children love variety and they thoroughly enjoy discovering answers for themselves. This enthusiasm is quite different from the unbridled exuberance that frequently characterizes field trips taken to totally unfamiliar surroundings where youngsters sometimes respond with frenetic physical activity or by bombarding the teacher with hundreds of questions.

In fact, teachers frequently avoid field trips because of their concern for the safety of the group or because of the ego-shattering impact of feeling ill-equipped to answer children's seemingly limitless questions.

If teachers regard the school ground as a three-dimensional reference library, many of their fears about going outside will be dispelled. We all use encyclopedias without knowing the total contents of the volumes or even paying attention to more than one article. The same thing should be true of teaching outdoors. A class going out to learn about the flying ash from an incinerator leaves the classroom with a definite goal. It has nothing to do with the names of the trees in front of the building, or the oxidation of copper on the rain spout, or any of the other observable phenomena.

So while the possible number of field trips is limitless (the ones described in this book are just samples), the teacher should deal with only one topic at a time.

And while it would be very hard to justify taking a one-hour bus trip to observe one tree for ten minutes, it is perfectly logical to leave the classroom for that length of time to focus in on one topic that is being discussed at the moment.

In fact, 10- and 15-minute trips are ideal for many topics. Others may be even shorter. A kindergarten or primary class studying shadows or an upper elementary class studying the sun may go out for five minutes at hourly intervals throughout the day to make observations. A class that returns to the building and discovers that their outdoor study has left some questions unanswered or raised new questions can easily return for further investigation.

Because they are short, school ground field trips can be a learning experience that enriches the curriculum even in the most tightly structured teaching situations.

Because they involve learning by observing, thinking, and doing, school ground field trips can bridge the reading and language barriers in situations where these are hurdles to learning.

Because activities carried out on the school grounds bridge the gap between abstract ideas and the real world, they make learning meaningful and often interest youngsters who are bored or have failed to relate to the classroom situation. This bridge-building between the classroom and the environment is tremendously important, not only because it gives purpose to classroom studies, but also because it gives purpose and direction to the trip.

Suppose primary grade children are studying about trees. Bark, branches, leaves, buds, flowers, and seeds can be just so many unfamiliar words to learn; but if the class goes outside and examines a tree and begins to talk about it, the words are discovered naturally. Back in the classroom, the children may make a big tree and add and label the parts, or every child could draw the tree, or the class may talk about the tree, or make a song about the tree.

If then the question arises, "Are all trees alike?", the class could go out another day and discover that the answer is both "Yes" and "No." "Yes," all trees have bark, leaves, a trunk, and branches. "No," some trees are bigger than others; some are older than others. The leaves on the trees along the street are different from the ones on the trees at the back of the play area. Some trees have needles instead of broad, flat leaves.

Back in the classroom, conclusions can be drawn, stories written, and charts made. After a very short time, children begin to notice trees in new places—at home, along the street, in a park, and in photographs in magazines and books. New ideas come up—ideas that can be tested, confirmed, or rejected on the basis of more trips. Each trip should have a follow-up session back in the classroom of recording, comparing, and fitting the pieces into the whole.

The children in one kindergarten in New York City who constantly used the school ground as a learning center were ready to enter first grade as readers at the end of the year. They were stimulated to learn by their interest and enthusiasm about the things they had dictated to their teacher for classroom charts after they had gone outside to learn.

A sixth-grade class in Massachusetts joined forces with a junior high school after they conducted a survey of school incinerator ash. Together they got more than 1,000 signatures on a petition which resulted in community action.

Several classes in a grade school in Kentucky that experimented with new vegetables in their school gardens influenced the type of crops produced by adults in the area.

Thus, the small area of the community that we know of as a school ground can be the basis for developing skills, attitudes, and concrete changes in the total environment.

Some classes may confine all their field trip activity to school ground studies; others may find that this micro-ecosystem raises questions that can best be answered by some longer trips. These trips away from school will be much more meaningful if they are based on many school ground explorations. A class that has had a lot of experiences with trees and animals will not be overpowered by a city park, a woods, or a botanical garden, but will approach this bigger area with the skills of observation and scientific inquiry developed and practiced in the school situation. There will be more questions to answer, possibly more materials to take along for recording, measuring, and collecting; but the concept of setting up goals and formulating questions before the trip, then returning to interpret data, will be well established, as will be the techniques of working in pairs, in small groups, or as individuals, and reporting back at a given time or signal.

A group of children who have done a survey of litter on their school ground and street and have decided that they need to visit the city disposal plant to get more information may not like what they see or smell, but they will be able to take the field trip without dramatic gagging and fussing because they are already familiar with the problem and have made the trip with definite goals in mind.

Just as the many school ground trips which follow are only samples which must be adapted and added to by the individual teacher, so the advantages listed above are also only samples of the types of things that can grow out of an exploration of the school ground by children working with teachers who are willing to tap one of the richest possible sources for environmental learning.

Of the Value of
Saying "I Don't Know"~~~~~~~~~~

Dr. Liberty Hyde Bailey, one of the greatest botanists of all time, dedicated one of his books to "a teacher who allowed a boy to grow." Frequently Dr. Bailey told about walking through the wild Michigan countryside of 1870 to a one-room country school and telling his teacher that he wanted to study nature. He then recounted how she looked at him and said, "Liberty, I don't know a thing about nature but we'll learn together. How many trees are there between your house and this school?"

When he answered that he didn't know, she said, "Liberty, that's the first thing you'll need to learn—to be observant." The following day when he came to school and told her how many trees there were, she said, "What kinds?"

All too often teachers are afraid to say "I don't know," or "We'll learn together." Yet this attitude is exactly what is necessary for really good teaching.

Children are born curious. My 14-month-old niece has a vocabulary of thirteen words and one phrase, "Whazzat?" It will only be a short time before "How" and "Why" are added.

"How" and "Why" are the basis for progress, scientific research, philosophy, indeed for humaneness. All normal people start life asking, "What—How—Why." But most of them get discouraged long before they reach high school age by adults who can't be bothered and ignore or even reprimand the questioner, or by well-meaning adults who feel they must tell all and leave nothing for the child to discover.

Children's natural curiosity is the most valuable asset a teacher can have. A child who is really interested in space will learn to read if there is enough material available on the topic. A child who wants to know how caterpillars transform to moths and butterflies will spend many hours caring for and ob-

7

serving caterpillars, cocoons, and chrysalids. Children who are encouraged in their curiosity and given the tools for finding answers for themselves will never be bored.

Youngsters who learn to ask questions, observe, set up possible answers, experiment, keep records, and think independently, will grow up finding life challenging and worthwhile. They will also have the ability to adapt to a changing world because they have learned skills that enable them to keep up with change, not facts that grow obsolete. This ability to adapt is tremendously important.

In the past years, I have frequently been asked how I can possibly be happy living in one of the most urbanized regions of the world when I grew up on a farm in rural Pennsylvania; or more particularly, what experiences in that very different habitat prepared me to be an urban environmental scientist.

In trying to answer that question, I have come to realize that I am keenly aware of, interested in, and concerned about the great interrelationships of this planet earth. The interrelationships between plants and animals and people, living and nonliving, earth forces and human forces, and our past, present, and future, exist everywhere. Names, numbers, or details may differ but the basics remain the same.

I started to learn this at an early age from my parents, who not only did not suppress my curiosity but who gave me tools for finding the answers to the questions I asked.

When as a preschooler I asked, "How do eggs get out of hens?" and my mother said she didn't know, she'd never seen a hen lay an egg, I suspect that she was hedging; but it offered a challenge to me. I spent many hours in the chicken house waiting to see what my mother had never seen.

Usually, a hen got in the nest, faced forward, sat still for a while, and finally left the nest cackling to announce that she'd deposited a fresh egg. But finally one memorable morning, a hen delayed at the feeding trays too long. She didn't have time to arrange the nest and face forward. She flew into the nest and the egg started to emerge. I watched fascinated, then rushed to lunch to regale the whole family with complete details down to the last straining and the final triumphant cackle; but my triumph was even greater than the hen's, for I, by myself, had made this wonderful discovery.

My father rarely said, "I don't know," but he consistently answered questions with questions. Often this worked well. Once in a while, the answer to his question was no more obvious to me than the answer to my own; and I can remember saying in anger, "If I knew the answer, I wouldn't have asked." Nonetheless, learning to break a big inquiry down into smaller questions is a tremendously important skill.

Dad tuned us in to the world around us with words like "look—watch—listen." One day when my brothers and I visited the farm across the street, the horses' watering trough was full of horse hair snakes. Mr. Gruber told us they had grown from horse hairs that had fallen in the water. When we rushed home to ask, "Do horse hairs really turn into snakes?" my mother said simply, "Why don't you try it and find out?" We not only tried horse hairs, we ran a constant series of experiments to find out.

From both my parents I learned to use books to help with answers. When the *National Geographic Magazine* arrived with an article on weeds, mother said, "Now we can find the names of all the weeds we pull in the garden." And we did, and enjoyed it no end because it gave us a way of communicating.

My father was interested in mushrooms, and when he discovered a new one, he made a spore print, then pulled out his copy of McIlvaine's *1000 American Fungi*. Together we studied the text and illustrations. I learned a lot about mushrooms but more important, I came to learn about books as places to go for information. I also learned about the importance of keeping records; for when we discovered a new mushroom, Dad recorded the time and place; and the next year he'd say, "We gathered oyster mushrooms on the dead log at the end of the field this time last year. Let's go see if we can find them again."

With this early background of discovering answers by observation and experimentation, learning to keep records, and using books as reference materials, I was hooked by the natural world at an early age and nothing could deflect me; but more important, I was well-equipped for rich living.

Today many children never have an opportunity to learn to discover their own answers, and gradually they lose interest in the "why" and "how" as they are fed other people's ideas during hours and hours of TV watching.

If schools are going to have a meaningful role in today's world, they must be more than dispensers of information and places to read; they must keep alive the natural spark of curiosity, they must nurture the ability to think, they must permit a child to grow.

Teachers with the vision and dedication to permit children to find out for themselves, to say, "Why don't you try it?" and to learn with the children, will find the school grounds an always-available natural laboratory, which can enrich any curriculum because it brings the real world into the classroom.

Organization
of this Book

This book has been written with the hope that it will help teachers to learn with their pupils and to discover the advantages of using the best of all teaching resources, their own school ground.

Obviously, a book of this type cannot and should not be complete. It does not attempt to give specific directions. What it does attempt to do is provide a listing of various school ground *Field Trip Possibilities*. Since every school ground is different and since all field trips should be a part of classroom experiences, this book can only suggest possibilities which the teacher can select and adapt as a starting point.

It is difficult to assign specific grade levels to the field trips because they are only suggestions designed to enhance, initiate, and give relevance to classroom learning. In general, it may be said that abstractions and complicated interrelationships require a kind of reasoning which has not developed in the primary grades.

Beyond this basic fact which makes some physical, Earth science, and ecological concepts more suitable for upper elementary grades, the way in which a field trip is used should depend on the background of the class and the topic being studied.

This kind of diversified learning growing out of a single field experience was dramatically demonstrated when the pupils in the training school at the State College at Fitchburg, Massachusetts tapped maple trees. The project grew out of a discussion in second grade. Some of the children had been to New Hampshire skiing and had seen sap buckets on maple trees. When they described their observations, a number of questions were raised about syrup and sugar production. Since two big sugar maples grew just outside the classroom window, and others grew nearby, it seemed that the best way to answer their questions about maple sap and maple syrup production was by first-hand experience. (See *Trees,* trip 15, p. 23.)

The children learned to identify the maples, selected the trees to be tapped, assembled the necessary equipment, and tapped the trees. They had an ample sap flow—more than they could handle—so they invited other grades to participate. They kept a record of the amount of sap they boiled down to make their syrup, and took responsibility for cutting off the sap flow with dowels when the project was complete.

The project took different forms in other classrooms. After the first-graders boiled down the syrup, they decided to have a pancake party with written invitations for invited guests. The fourth-graders had a party, too, a sugaring-off party reminiscent of bygone days. They took their thickened syrup outdoors and poured it on clean snow for a special and unusual treat.

Children in the other grades ate their syrup on crackers and turned their attention to various topics suggested by their experience. For fifth-graders, the syrup production initiated a study of colonial home crafts like natural dyes and candle-making; for third-graders, it led to a study of foods of the American Indians; and for sixth-graders, it raised questions on boiling temperatures of different liquids, rate of evaporation, surface tension, specific gravity, and the use of hydrometers.

In other words, one school ground activity resulted in six different projects reflecting the classes' course of study, the children's maturity, the children's interests, and the teachers' imagination. Obviously, some of these projects could be interchanged. Anyone can enjoy making pancakes and having a party. The first-graders used a prepared mix and the teacher worked out a recipe that they could successfully read and measure. What was a reading and simple measurement experience for first grade could become a practical arithmetic project for older children. As they decided how many times they would multiply or divide the recipe to make the desired number of pancakes, they could plan and do the necessary shopping.

On the other hand, the physical science concepts explored by sixth grade could only be studied by youngsters with advanced arithmetical skills and knowledge.

This same diversity of approach is possible with any of the field trips in this book. When you and your students start exploring the environment together, you will find you are sharing a learning process that enriches and gives meaning to classroom instruction while it opens doors to all kinds of new experiences, which can never be regimented to a rigid course of study.

Just as field trips should not be an isolated part of classroom teaching, they are supported here by three types of material. Each subject is introduced with a page or two of background information on the general topic.

Next, some *Related Classroom Activities* are suggested. These are only a few of the almost limitless possibilities. They are included as a reminder that field trips need an introduction and a follow-up in the classroom; however, the exact introduction and follow-up belongs to you and your class. Frequently, a follow-up for one trip can be the introduction to the next trip.

There is also a section on *Teacher Preparation*. It is absolutely essential for teachers to spend some time on the school ground, locating the area that will be used and deciding on the best approach. After you have used the grounds for a number of years, the time spent this way may be minimal; but it is important to know that the tree you used last year is still alive; that the repairs on the building did not destroy your best teaching site for chemical change; that no temporary hazards exist.

A *Cross-Referenced Listing of Field Trips for Hard-Topped School Grounds* is given at the back of the book. These trips can be conducted on a school ground without an inch of exposed soil. There is also a listing of trips which can be taken on completely hard-topped school grounds with plantings only in concrete cracks and broken concrete areas.

Teachers can obtain additional information on all topics in the book by turning to the annotated list of *Supplementary Materials*. It includes not only books but also recordings, project materials, guides, and sources of supplies. Reading levels, as well as the subjects for which each material can be used, are also indicated.

The index has been carefully cross-referenced, for in this world of inter-relationships, a concept listed under geology might also be meaningful in plant study, or a term relating to seeds might be significant in an animal study, and so on *ad infinitum*. The index will also help with unfamiliar terms since these are defined only on their first appearance in the text.

Finally, the idea for this book grew out of in-service training courses taught by the author and organized by Hannah Williams at Wave Hill Center for Environmental Studies in the Bronx, NY. Through Mrs. Williams' efforts and bridge-building between Wave Hill and the schools of the Bronx, it was possible to conduct these courses on the school sites of the participants and on city streets as well as at the city park at Wave Hill.

In the courses, teachers learned to use the outdoors to supplement their classroom studies. When one young man compared his experiences taking his class out on an unstructured field trip to that of taking his class to exactly the same area to examine a rock outcropping after they had read about joints in a geology lesson, the idea of doing a survey of possible short field trips on the school ground was born. He said, "The first trip was just horrible. I took the students out and they asked a hundred unrelated questions and I couldn't answer any of them. I was never going out again. But the geology trip was different. The students raised questions in the classroom. We went out to the rock out-cropping and they discovered the answers for themselves. We came back and their textbook material now made sense. It was a great experience."

After the teachers in several courses had made school ground surveys and experimented with school ground teaching, the Bureau of Audio-Visual Aids of the Board of Education of the City of New York provided a photographer to take pictures of Barbara Alweis, one of the teachers in the in-service course, con-ducting field trips with her kindergarten class on the P.S. 51 school ground. These photographs were made into a film strip with an accompanying audio-tape and offered for sale to schools and colleges with teacher training programs. A manual to accompany the strip was published by Wave Hill Center for Environmental Studies. The material in the manual was tried in classrooms throughout the United States both with and without the filmstrip.

Development of this manual into a full-length book has been made possible by the assistance and cooperation of and the exchange of ideas with Richard Whittemore, Betsy Weiss, Dorothy Dunbar, and Marian Manfredi of the J. G. Ferguson Publishing Company, a subsidiary of Doubleday & Company, Inc. In addition, my husband, Robert S. Russell, has served as an always-available sounding board and helpful critic.

The names and locations of the teachers and others who wrote evaluations and made recommendations after trying the material in the manual will be found below. Their contribution and encouragement was invaluable in determining the direction of the book. Barbara Alweis, Teacher, Bronx, NY; John Brainerd, Professor, Springfield College, Springfield, MA; Norma Brodsky, Teacher, New York, NY; Mary Civitielo, Teacher and Science Coordinator, Long Island City, NY; Doreen Cleland, Professor, College of Education, Kingston Vale, London, England; Ted Gaskin, Science Coordinator, Scarsdale, NY; Jerry Gentry, Director, Editorial Research, Ambassador College, Pasadena, CA; John A. Gustafson, Chairman, Biology Department, Homer, NY; Ruth E. Hartley, Chairman, Department of Growth and Development, University of Wisconsin, Green Bay, WI; Eleanor Horwitz, Teacher, Eugene, OR; Ruth R. Howell, Early Childhood Education, City College of New York, New York, NY; Lorraine B. Ide, Elementary Science Supervisor, Springfield, MA; Gladys Kleinman, Professor, Science Education, State College, Jersey City, NJ; John A. Kossey, Professor, Ambassador College, Pasadena, CA; Frances Lamey, Teacher, Lexington, MA; Warren M. Little, Coordinator, Conservation Education Center, Massachusetts Audubon Society, Lincoln, MA; Yvette Marrin, Assistant Director, Marble Hill Nursery School, Bronx, NY; Anna Mikaelian, Teacher, Fitchburg, MA; Mary E. Montgomery, Teacher, Oberlin, OH; Gilbert Mouser, Professor, Michigan State University, Lansing, MI; Dorothea Mulaik, President, Western Division, American Nature Study Society, Salt Lake City, UT; Carl H. Nunley, Outdoor Education Program Consultant, Allentown, PA; Lorraine B. Quesada, Teacher, Townsend, MA; Joan Quinlan, Teacher, Lunenburg, MA; Robert Rice, Science Coordinator, Bronx, NY; Gladys S. Richard, Teacher, New York, NY; Marlene Robinson, Teacher, Bronx, NY; Joan Rosner, Science Coordinator, East Elmhurst, NY; Ruth F. Ryan, Teacher, Manchester-by-the-Sea, MA; Alfred Satsuk, Teacher, Bronx, NY; Milton A. Stein, Teacher (retired), Brooklyn, NY; Eve Strickler, Member, Conservation Education Committee, Sierra Club, New York, NY; Phyllis Tandlich, Teacher, Bronx, NY; Jacques Van Pelt, Outdoor Education Project, Inuvik, Canada; Jack C. Wood, Director, Audubon Farm, Dayton, OH.

A study of trees on the school ground can not only span the seasons but can provide an opportunity for young people to observe growth, be responsible for maintenance, and study interrelationships over several years—thus providing a base for environmental and ecological understanding.

Plants

The role of plants becomes very dramatic if we contrast the moon with the Earth. We say the surface of the moon differs from the Earth because it has neither air nor water. This is true. But even if it had air and water but no plants, it would be a barren, craggy wasteland, just as the Earth was until green plants first emerged from the oceans. Plants began to make soil by hastening the breakdown of rocks by both physical and chemical means, by adding organic material to the rock debris, and by anchoring the shifting rock fragments against the action of air and water.

Of course, soil-making and holding are only one of several life-sustaining activities performed by plants. Green plants either directly or indirectly serve as the source of food for all other life, and in producing surplus food they constantly replenish the world's oxygen supply. In fact, there is a great deal of concern among ecologists today about the wholesale destruction of plant life, on land by careless and wasteful resource use, and in the ocean through the runoff of modern chemicals like detergents, DDT, defoliants, and oil spills.

Nongreen plants are just as essential to life as green ones. Several kinds of bacteria are responsible for taking pure nitrogen from the air and combining it with oxygen to form the chemicals that are needed for protein production.

Thousands of other kinds of fungi (nongreen plants) ranging from bacteria to mushrooms are members of the world's cleanup crew. In living on dead plant and animal materials, they break big molecules down to small ones, which are then reused by plants in the production of food and oxygen.

If it were not for the work of these decomposers, the world today would be a lifeless globe cluttered with the dead bodies of ancient plants and animals; and no building blocks would be left for modern plants and animals.

15

While some of these relationships can only be observed in chemical laboratories, others can be seen on even the most barren school grounds.

RELATED CLASSROOM ACTIVITIES

A study of plants could grow out of many things: the study of pollution, food, food chains, adaptation, plant products, or interrelationships. Or it could start with a potted plant brought to a classroom, or a bouquet of flowers, or the first dandelion growing in a pavement crack, or a crocus pushing up through the snow.

Regardless of the starting point, sooner or later the class will arrive at a point where actual experience with plants will make the learning more meaningful, where it is logical to say, "Let's go outdoors and see—where—what—how."

After you've been outdoors and have come back to summarize, record, and discuss the students' observations, there will probably be new questions. These may involve taking another field trip, devising an experiment, or doing some book research. They may even include growing some plants in a garden out on the school grounds or on the windowsill, or, in a very dry classroom, in a terrarium.

A terrarium is really a miniature greenhouse. It can be made from a clear plastic breadbox or hat box or from an empty jar. Straight-sided, wide-mouthed, half-pint jars of the type in which pickles and peanut butter are sometimes sold are excellent for small terrariums. Empty gallon jars, either standing upright or on their sides, make good larger terrariums.

Regardless of size, the terrarium should have a layer of pebbles on the bottom for drainage. If charcoal is available, a few pieces help control mold. Then the pebbles should be covered with an inch or two of potting soil. Seeds, plants, or cuttings can be added, the soil watered, and the lid put on. Unless the miniature greenhouse is being used as a temporary growing site for seedlings or cuttings that will be transplanted later, the size of the container will determine the type of plants that can be grown.

Check the top daily for moisture. If a heavy coat of water appears on the top and the soil is wet, remove the lid and permit some water to evaporate. If no droplets form on the top and the soil is dry, you may need to add a bit more moisture.

Caring for plants and observing plant growth in these small greenhouses can help youngsters develop new understandings and appreciations. It should also help older children gain new insights into the seriousness of pollution and the use of defoliants and perhaps lead them to take new responsibilities in the care of all plants, including street trees and school and city plantings.

TEACHER PREPARATION

The school grounds should be inspected ahead of time for any field trip and the site or sites selected for the desired research. At the same time the teacher should make some decisions about class organization. Are there enough sites for children to work in pairs? In groups of four? If they work in groups of eight or ten, will everyone be able to see and participate? If several trips are being made, it may be necessary to devise some technique for marking boundaries with stakes, string, or paint.

FIELD TRIP POSSIBILITIES

1. Examine a potentially barren area like a concrete play area, a walk, stone or concrete steps, a part of a stone wall, a building or a rock outcropping, for plants. Can you find algae looking like a green coat of paint on rock surfaces, or lichens looking like black, yellow, brown, orange, gray, or grayish-green splotches? Can you find moss growing in sidewalk cracks or between stones? Are there weeds or grass growing in sidewalk cracks or places against buildings where broken fragments of concrete have collected? Are bigger plants like young Ailanthus trees, elm or Norway maples trying to become established? From your discoveries, can you draw some conclusions about plants and soil formation? About the soil requirements of plants?

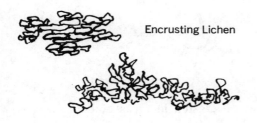
Encrusting Lichen

Foliose Lichen

Pixie Cup Lichen

Lichens come in many colors and shapes. Some, like encrusting and foliose lichens, look like peeling paint or lace doilies, respectively. Others have reproductive structures that make identification easy, like the pixie cup lichen.

2. If by chance the area is totally barren, this also provides a basis for plant-soil relationship discoveries. While algae and lichens can grow on rock surfaces without any soil, algae must have moisture in the air and lichens will not grow in areas of high sulfur or carbon monoxide pollution. Mosses, ferns, grasses, weeds, and trees have progressively greater soil and water requirements.

3. Do a survey of all the kinds of plants on the school property. There is no need to identify the plants by name. The list can say: four kinds of grass, two kinds of mushrooms, five kinds of trees. Drawings or leaf, twig, or flower samples mounted on a chart will demonstrate the different kinds.

4. If there is a hedge on the property, can you see evidence of soil-making in the form of leaves caught and held by the plants, and of soil-making and soil-holding by the difference in ground level beneath the hedge and beyond it?

5. Dig up a cube of soil in an area where plants are growing and another from an area where the ground is bare. Compare them in terms of hardness, moisture, color, and animal life. Can you see other differences?

6. Find white clover plants growing in the lawn. Have children pull them. Notice how the plant is anchored to the soil by hundreds of roots that grow wherever the stem touches the soil. Find the small spheres on the roots that hold the bacteria that produce nitrogen compounds. Discuss the role of clover in soil enrichment and soil-holding.

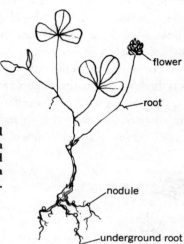

White Clover

White clover is one of the best plants for soil building and enrichment. The small roots which arise wherever the plant touches the ground make a soil-holding network. The nodules contain bacteria which produce nitrogen compounds.

7. Go out during or after a rain and observe soil erosion in areas of bare soil as compared to areas covered by grass or other kinds of vegetation.

8. Make small cages of fine wire screening. Put bits of meat or fruit or leaves in each cage and fasten it outdoors on a sheet of sticky flypaper so small insects cannot approach from above or below. How long does it take for decomposition to begin? What causes the decomposition? How do bacteria and other things get to the food?

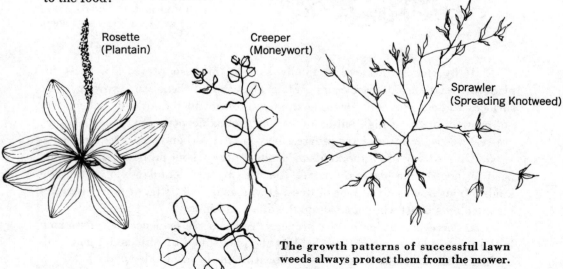

Rosette
(Plantain)

Creeper
(Moneywort)

Sprawler
(Spreading Knotweed)

The growth patterns of successful lawn weeds always protect them from the mower.

9. Take a census of lawn weeds before the grass is mowed and afterward. What forms do successful lawn weeds take? Try to find several examples, of rosettes, creepers, sprawlers.

10. After trip 3, make a map of the school ground showing where different kinds and numbers of plants are growing. Correlate this with soil, sunlight, wind, and any other factors which affect plant growth.

Trees

Trees are the largest of all land plants. Like other green plants, they have leaves and produce flowers and fruit, but they have special structures that go with bigness. The woody stem we call the trunk belongs to trees alone. It is covered by the tough bark that protects the living tissue beneath it from drying out and from bacteria and animal enemies (much like human skin). As the tree trunk grows, the bark peels and folds and wrinkles in ways that are distinctive for different kinds of trees. Compare a London plane tree to a Norway maple in this respect.

When bark is broken or damaged, the tree grows new bark to heal the wound. Often this is a different texture and color. Children can locate these bark scars, which, like scars on their own skin, tell of accidents and injuries.

On many trees, they also will be able to find circular scars where shaded branches died and broke off.

When injuries are too great, a tree may not be able to cover the exposed spot before bacteria and insects get into the wood and cause decay. Concerned tree owners frequently cover injuries with paint, tar, or concrete.

London Plane Bark

The peeling varicolored bark of the London plane tree makes it an easy city tree to identify. Examine the trunk and correlate the different colors with the relative age of the pieces of bark.

The leaves of the tree are the tree's food factories. For every molecule of surplus food produced and stored in the trunk, roots, seeds, or fruits, the tree releases six molecules of oxygen. Thus, a healthy tree makes oxygen available.

The thousands of leaves are arranged in such a way that some sunlight hits every leaf. The buds form in the summer ready to produce leaves and flowers the following spring. Powerful roots both support the tree and carry water for food production. Finally, most trees produce millions of seeds, all provided with some technique for dispersal and planting.

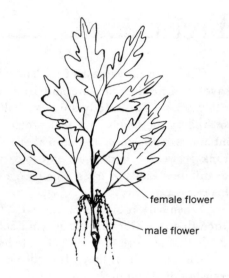

female flower

male flower

White Oak

The egg contained in the female flower of the white oak will grow into an acorn after it has been fertilized by sperm from the pollen of the male flower.

Tree flowers, seeds, and fruits all have their seasons. Some trees, like buckeye, catalpa, apple, and cherry, have showy blossoms. Many others have flowers that are equally beautiful but go unnoticed because of their small size. Discovering maple, oak, and other little-known tree flowers can be an interesting spring activity.

flower

fruit

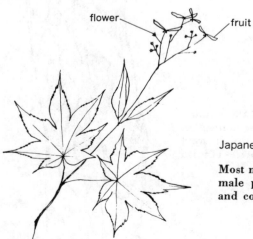

Japanese Maple

Most maple flowers contain both male and female parts. While the flowers vary in size and color, all produce paired, winged fruits.

RELATED CLASSROOM ACTIVITIES

The study of trees may grow out of many things: a story about a tree, a study of wood or wood products like paper, a picture of a tree, a colored leaf brought to the class by a child, a trip outdoors to draw a tree in an art class, a discussion of plants, or the discovery that trees and other plants help fight pollution. The decision to go outside and look at a tree could also grow out of a vocabulary study. Conversely, a vocabulary study could grow out of an examination of a tree.

Children who have looked at trees on the school ground may be asked to examine the trees along the street as they go home from school. Their reports may raise new questions which can only be answered by going out and looking at more trees.

Leaves can be gathered on a field trip, placed between pages of a pulp magazine or an old telephone book and dried, then mounted between sheets of clear plastic or placed between two pieces of wax paper and ironed with a cool iron to make window transparencies.

Or the leaves may be used to make leaf rubbings and leaf prints. Of course, making leaf prints can be another way to introduce trees and provide a springboard for further study and field trips.

A chart summarizing measurements and discoveries makes a fine follow-up. It also may serve as an invitation to more trips to discover how the tree changes with the seasons.

TEACHER PREPARATION

Trees are dependable. Once you begin to be acquainted with the tree on the corner, you will find that you have established a relationship with a very interesting living thing which will remain on that same corner day after day, week after week, year after year. You'll soon discover many teaching possibilities that you can build around it.

There is no need for you to know its name. If children ask the tree's name, you can suggest that after they have made a chart of characteristics they can look it up. Even if you do know the name, this is the best answer you can give youngsters. Learning should not be memorizing but discovery.

Pruning shears or a pocket knife should always be used in collecting twigs so that jagged ends and torn bark do not disfigure the tree and leave it open to disease. Children should be taught that indiscriminate gathering from lower limbs deforms the tree and spoils everyone's enjoyment. The teacher's example is the best way to accomplish this.

FIELD TRIP POSSIBILITIES

1. Get acquainted with one tree. Admire its trunk. Notice the branches, twigs, and leaves.

The bigness of a tree gives it a special quality. Children enjoy measuring a tree with hands and arms. How big around is the trunk? Can you put your hand around it? Two hands? Your arms? Does it take more than one child to reach around it?

Examine and feel the bark. Look at its ridges and patterns. Feel its texture and hardness. A bark rubbing can be made with a wax crayon and paper as a record of the tree.

2. Select a tree to visit repeatedly. Observe and record changes. A dated chart with entries such as mounted leaves, twigs, flowers, rubbings, fruit, and/or seeds will enable children to make seasonal comparisons. Written summaries can stimulate further reading.

3. If there are several trees of the same kind on your grounds, compare them for size, shape, time of leaf coloration, and the time when buds open. If there are differences, can you think of explanations?

4. Make a trip to see how many different trees you can find on your grounds or street. You can make a census without listing names by mounting things like a sample leaf, twig, fruit, and/or seed from each kind of tree. It is good to have books available for children who may enjoy discovering names by making a comparison of the specimens with illustrations and printed material.

5. Go outside and play a game with tree shapes. Let someone in the middle of a circle represent one of the trees by using his arms as branches to show the shape of the crown. The first child that points to the growing tree being represented moves to the center. Or the class can be divided into two lines with all the children in one line representing the same tree and the ones in the other guessing.

6. Measure a small tree before buds open and again in June to show how much upward growth has taken place.

7. Look for bark damage. Is there anything you can do to protect trees on your school grounds? Can your class initiate an educational program?

8. If you are studying street trees, what things have been done to protect roots and bark? To provide water?

9. Explore relationships of trees and automobiles. Are trees at bus stops as healthy as others on the street? Can you relate this to air pollution? Can you find other damage from cars? From people?

10. Make a study of tree flowers in the spring.

11. Tree seeds ripen at different times. Some, like elm, willow, poplar, and silver and red maple, ripen in late spring or early summer. Many others ripen in the fall. Still others, like black oak, pine, and spruce, take two or three years to ripen. Many tree seeds are dispersed in winter. Some go sliding over the snow. Field trips to study tree seeds can be just as much a seasonal study as tree leaves are.

12. In the spring, take a field trip to observe tree seedlings. Compare the growth of a tree seedling to the growth of a lima bean. Dig up some seedlings and pot them for classroom observation. If seedlings are abundant, do a census once a week. What happens to many of the seedlings? Are these population controls important?

13. Go outside in a light rain and notice the effect of the tree in breaking the force of raindrops as water drips from branch to branch and finally runs down the trunk to the ground. Imagine the effect of a "roof of trees" like a forest.

14. If your school ground hasn't a tree, why not plant one? Even an all-concrete school ground could have a tree in a large container. School grounds with trees might be improved by adding a new kind.

15. Any school that has maple trees on its grounds can experiment with producing maple syrup. Although sugar maple trees and occasionally red maple trees are the ones used commercially, any maple tree can be tapped. This includes the more common Norway, ash-leaved, and silver maples of the city; the English sycamore maple and sugar maple of the South; and the big-leaf maple of the Northwest. The Indians also tapped birch trees and used the sap to make sugar.

To tap a tree, bore a 5- to 10-centimeter deep hole with a 10-millimeter bit about one meter above the ground. Drive a spile (a spout) halfway into the hole. In sugaring country, spiles can be purchased in a hardware store, but a 20-centimeter length of half-inch copper tubing will do as well; if elderberry bushes are available, an Indian spout can be made by cutting a 20-centimeter section of a stem and pushing out the soft pith in the middle. A bucket for collecting can be made by adding a wire handle to a number 10 tin can.

Collecting time extends from the beginning of sap flow (usually February in New England and earlier farther south), until the trees are in full leaf. It is best when nights are cold and days are bright and sunny.

As much as 3 liters of sap can be collected from one hole in a day's time, and big trees can have several holes. It takes 25 to 35 liters of sap to make a liter of syrup.

The sap can be boiled down in open pans on an electric burner in the classroom. (Never boil sap in rooms with wallpaper.) This is a several-day process. Children can take turns watching the sap boil and adding more as it boils down. The process can be terminated when school is dismissed and be resumed the next morning.

As the sap thickens, it tends to boil up and overflow the pan. This can be checked by adding more sap, or if the desired quantity of syrup is being approached, by dropping a small amount of butter on the surface. Sap should be refrigerated over the weekend.

When enough sap has been collected to make the desired amount of syrup, the spiles should be pulled out and the holes sealed by driving 2-centimeter wood dowels into them and cutting the dowels off flush with the bark. In a year or two, only a small circular design in the bark looking exactly like a branch scar will be visible as a reminder of the sugaring activity.

Maple syrup and sugar production can relate to many topics. When the children of Manhattan Country School made syrup, they collected, duplicated, and sold a cookbook of maple recipes. (For additional activities concerning tapping maple trees, see *Organization of This Book*.)

Sap may also be collected by snipping off the end of a branch and attaching a bottle or other container to the branch in such a way that the sap drips out of the wound into the container. Several bottles of sap can be collected very quickly in this way, tasted, and then boiled down to produce enough syrup so everyone can have a sample and compare it to the original sap.

Leaf Coloration

Children and adults alike are intrigued by the beauty and drama of autumn leaf coloration. The phenomenon has been the subject of Indian legends, poetry, and much scientific research. Fortunately, the research has not spoiled the poetry or joy of appreciating the rich coloring of a beautiful fall day. This is partly due to the fact that there are no firm, inflexible explanations. Leaf coloration is the result of a combination of factors, and some of these factors vary from year to year.

Leaves turn yellow when the chlorophyll breaks down and disappears from the leaf and the yellow pigments that were already there become visible. Almost all leaves have yellow pigments, but during the summer months the darker chlorophyll keeps them hidden.

Many of the deep reds in plants like Schwedler's maple, red-leaved barberry, copper beech, and some Japanese maples also become brighter when the chlorophyll disintegrates. The red pigment existed in these plants all along but was somewhat muted or even looked brown when covered or mixed with the green chlorophyll.

On the other hand, colors like purple, pink, scarlet, and many other shades of red are newly formed in leaves in the fall of the year. They come from chemical change. At the same time that chlorophyll disintegrates, colorless sugars, gums, and tannins in some leaves undergo chemical changes and become brightly colored compounds.

Both the disintegration of chlorophyll and the chemical changes in the compounds in the leaf are triggered by the change in length of day and night, but the extent of chemical change and the brillance of the resulting colors are related to temperature, moisture, and the amount of bright sunlight.

The brightest coloration exists in years when there has been adequate rainfall during summer and early fall and when fall nights are cold and days are bright and sunny.

Both the ability of leaves to change color and the potential color range are inherited characteristics. Regardless of length of day or night, moisture, temperature and sunlight, pine trees and most other conifers will always remain green, as will most kinds of holly, laurel, and grass.

Just as some people can never develop a suntan because they have not inherited the gene that initiates this chemical activity, so many leaves never develop red coloring because they have not inherited the gene that is responsible for producing the chemicals that form these pigments.

This is true of the four trees most commonly seen on the city streets: Ailanthus, London plane, Gingko, and Norway maple. In addition, these trees rarely reach beautiful shades of yellow because of the effects of pollution, and other trees that normally turn red, like scarlet oak, frequently fail to do so in polluted areas.

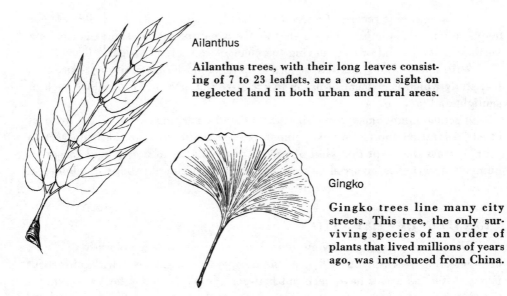

Ailanthus

Ailanthus trees, with their long leaves consisting of 7 to 23 leaflets, are a common sight on
neglected land in both urban and rural areas.

Gingko

Gingko trees line many city
streets. This tree, the only surviving species of an order of
plants that lived millions of years
ago, was introduced from China.

Some pollutants cause damage by coating the leaves with dirt and grease, while others enter the leaves and interfere with chemical processes. The best leaf coloration is found in areas with fresh air and a cool climate.

Of all trees, sugar maple is the most brilliantly colored. As the chlorophyll disintegrates in its leaves, the rich yellow hues become visible and at the same time the stored sugars begin to form brilliant red pigments. The resulting red, orange, and yellow make the tree look like a living flame.

Ash trees with their purple hues are also distinctive. Biochemists have discovered that red pigments form only when the plant sap is acid, and blues form when the plant sap is basic. Colored leaves might be considered giant ribbons of litmus paper.

While we usually think of colored leaves in connection with trees, the leaves of many shrubs, vines, and bushes also change color, as do some weeds and even an occasional grass.

RELATED CLASSROOM ACTIVITIES

Colored leaves pressed between clear contact paper or wax paper make beautiful transparencies and serve as a delightful reminder of a collecting trip or an interesting introduction to the study of leaf coloration.

In an art class, colored leaves may be used to make collages by cutting up the leaves and gluing them with rubber cement onto paper or light cardboard in exciting patterns.

Concepts of the way in which leaf coloration changes with the removal of chlorophyll can be worked out with water-color paints or colored overlays in an overhead projector. Start with yellow—add green, what happens? If you start with red and add green, what color is the leaf? When you remove the green, what happens?

A photographic record of a tree that changes color dramatically like a sugar maple might be made by taking a shot of the same tree at the same spot once a week over the period of time during the change from green to leaf fall.

With upper-grade children, record-keeping could be an important part of a leaf study which is concentrating on the relative importance of variables such as sunlight and moisture.

A school might amass some significant data by keeping careful, dated records of leaf coloration and factors like temperature, sunlight, and rainfall from year to year. A class that kept this kind of record would begin to understand the great number of variables involved as well as some of the complications of biological research.

TEACHER PREPARATION

Watch the area around your school for changes in leaf coloration. If you are collecting leaves, you will need newspapers or magazines with absorbent pages. After the leaves have been put between the pages, put a heavy weight on them to keep the leaves from wrinkling. Keep them pressed for several days. Leaves dried this way can be used for a long time. They can even be "reconstituted" in winter months for making leaf prints by putting them between moist newspaper for 24 hours.

FIELD TRIP POSSIBILITIES

1. Go outside and make a collection of colored leaves representing the variety of hues visible on your grounds. These leaves may be dried between newspapers and mounted with rubber cement or used to make collages, transparencies, or leaf prints and rubbings.

2. Find two trees of the same kind and compare them. If there is a difference in color, hue, date of leaf change or leaf fall, can you find any possible explanation? Compare the available moisture, age of the tree, the health of the tree, and the amount of daylight.

3. A vine whose branches grow on two walls facing in different directions can dramatize the role of light in leaf coloration, since you are dealing with only one plant having the same rainfall, soil, length of day and night, and basic weather conditions. Make a record of the time coloration starts on each side, the amount of coloration, and the time leaves fall, and correlate them with the hours of sunlight. What conclusions can be drawn?

4. Sometimes the two sides of a tree will be dramatically different in color. Can you find an explanation? Is one side shaded? Exposed to a street light?

5. Azaleas, barberries, and other bushes with overlapping leaves often have "leaf photographs" on their lower leaves. If you lift upper leaves, you will find them "pictured" in green or yellow and surrounded by a red background on the leaf that they shade. This concept of photographs (light drawings) is very exciting to children.

6. When nights grow longer, the tree begins to lay down a corky layer between the twig and the leaf which cuts off sap movement and initiates the

chemical changes. If sumac trees are available, you can experiment with cutting off circulation before the tree does it by breaking the midvein of some of the leaflets and observing the changes.

7. Examine a weed patch for leaf coloration.

8. If you find reds and blues among the leaf colors in the weed patch, try to extract the sap by crushing the stems and test it with litmus paper.

9. Compare the red of leaves like Schwedler's maple and copper beech, which contain red throughout the growing season, to the reds of chemical change in trees such as gum, sugar and red maple, and scarlet oak.

10. Compare the leaf coloration of a street tree to one of the same kind on a lawn. Is there any difference between two trees of the same kind along the street, one by a bus stop and one at a distance from it?

Buds

As soon as leaves are fully expanded and producing food, woody plants begin to make buds for the following year. At first the buds are small and green, but they can easily be found in the leaf axils by late spring.

As the summer wears on, they increase in size. Finally, they harden and go into a resting stage which lasts throughout the winter.

The shape, color, form, and composition of buds is just as much an identifying characteristic of a woody plant as its leaves.

Buds contain the rudiments of branches with leaves, flowers, or both leaves and flowers. In plants that bloom in early spring, the flower bud is frequently larger and a different shape than the leaf buds. For instance, a magnolia tree has its flowers all prepared and ready to open when warm weather arrives. A census of the big flower buds can be used to predict the potential bloom for the next year. There can't be more flowers than those already prepared. Of course there can be fewer, depending on factors like temperature, wind, ice storms, and the things people do.

Many buds are enclosed in scales which protect the tender growing parts from mechanical injury and from drying out. Frequently, these scales are coated with waxes, gums, or resins, or covered with hairs.

A few woody plants produce buds without scales. These are called naked buds. The buds of pignut hickory and many viburnums are examples of naked buds. Leaves of these are thickened, rolled in from the edges and covered with hairs.

Viburnum

In this naked bud, stiff hairs protect the folded viburnum leaves from drying out.

The packaging of leaves in a bud is very interesting. Maple leaves are folded along each radiating vein like a fan. Beech leaves are packaged flat. Tulip tree leaves are folded in half on the midvein, then folded forward on the petiole (leaf stem).

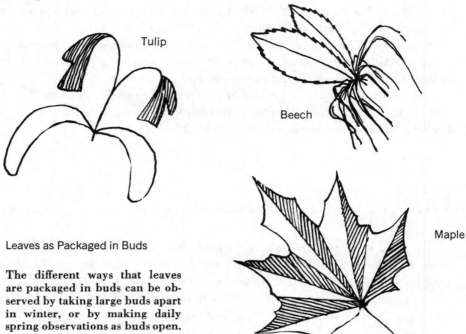

Tulip

Beech

Maple

Leaves as Packaged in Buds

The different ways that leaves are packaged in buds can be observed by taking large buds apart in winter, or by making daily spring observations as buds open.

In areas with cold winters, many kinds of woody plant buds must be frozen before they open. Sometimes when a warm Indian summer follows a period of freezing weather, some flowers of early-blooming plants like azaleas and forsythias open, since their built-in timer is set for blooming in warm weather following freezing weather.

RELATED CLASSROOM ACTIVITIES

Opening buds is an excellent way for children to learn about them. While all buds contain the rudiments of twigs and leaves or flowers, only large ones will be well enough developed for children to observe. Horse-chestnut, beech, and tulip buds are three good types for classroom use.

Forcing buds to open by placing them on the windowsill is another good introductory or follow-up technique. Some buds must be frozen before opening, so, unless you are conducting an experiment to discover which ones require freezing, it is better to bring twigs into the classroom in January or February. Again, the larger buds open better, but children can learn from having a variety and drawing their own conclusions.

Any sustained study of buds will involve keeping records and making charts.

TEACHER PREPARATION

Examine your school area for woody plants. If they grow behind a locked fence, arrange for the custodian to open the area for your class. Always use pruning shears or a knife when collecting twigs. Collect no more than one twig from each plant in landscaped areas.

If you are collecting twigs for your classroom, get them from a woods or from a wild tree. Arrange for clippings to be saved for you at a botanical garden or when an orchard is being pruned. Obtain twigs and small branches of evergreens at a place where Christmas trees are being sold.

FIELD TRIP POSSIBILITIES

1. In the fall of the year, examine trees and bushes on your school grounds. Can you find buds on all of them? When did the buds form?

2. Notice that buds always grow terminally (at the end of the branch) and in leaf axils (the place where the leaf petiole joins the branch). A few trees like sycamore, London plane tree, and yellow-wood have leaf petioles which completely surround the bud. Discovering these hidden buds can be great fun in the fall when the leaves are still on the tree but are turning yellow and pull off easily.

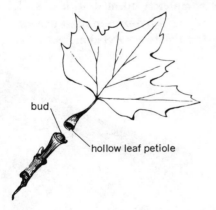

bud

hollow leaf petiole

London Plane

Discovering that next year's growth is formed and packaged in buds during the previous summer is always surprising. This surprise is doubled when examining trees like the London plane, where the bud is hidden in the leaf petiole.

Even when leaves have fallen, this relationship of the bud to the leaf of the year before can easily be observed, for every leaf makes a scar on the twig. The size and shape of the leaf scar tells about the size and shape of the petiole.

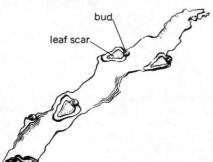

bud

leaf scar

Ailanthus Twig

When the large leaves of the Ailanthus tree fall, they make big scars that are easy to examine. The dots in the leaf scar are the vascular bundles—the bridges where leaf veins were connected to the branch providing for a flow of sap and other materials back and forth from the leaf to the woody parts. The buds which have formed in the leaf axil will be the beginning of next year's branches.

3. If you have azaleas, dogwood, magnolias, or rhododendrons on the school grounds, see if you can discover two kinds of buds. The large buds of these plants are flower buds. How many flowers or flower clusters may you have next spring? If you make a count in the fall, record your finding. Repeat the trip in the spring. Did the tree or bush develop its full potential? If not, why not?

4. Select one woody plant. Visit it and record its bud story the first day of fall, winter, spring, and summer.

5. Make a chart of the different woody plants on your grounds. Either draw, photograph, or mount specimens of the different types of buds on each plant. Keep a record of the dates of the opening of flower and leaf buds. Compare the size of flower buds that opened before the leaves to those that opened with the leaves and those that opened later.

Seeds

Seeds are immature plants. This is commonly illustrated by taking a soaked lima bean apart and examining the two seed leaves (cotyledons) surrounding the immature plant (embryo). This small plant is completely formed with a stalk and the first two true leaves. It can be seen equally well by examining a peanut. The papery seed coat is easily removed disclosing the two cotyledons with the tiny peanut plant between.

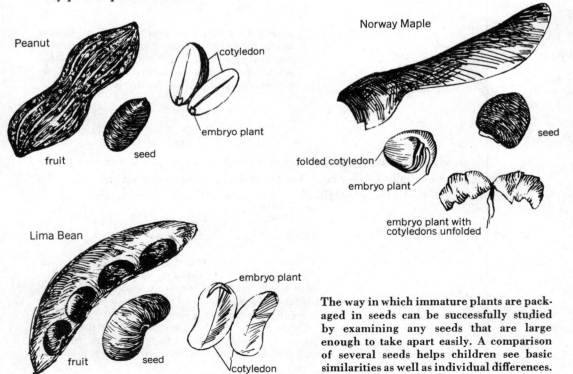

The way in which immature plants are packaged in seeds can be successfully studied by examining any seeds that are large enough to take apart easily. A comparison of several seeds helps children see basic similarities as well as individual differences.

Both peanuts and beans have large cotyledons which are packaged flat. Norway maple seeds, on the other hand, have their strap-shaped cotyledons folded to fit into a small area (just as you fold a belt before wrapping it). The way in which it is folded can be appreciated by gathering Norway maple seeds in the fall, winter, or early spring and putting them to soak in water for 24 hours. After that, the seed coat can be peeled off and the cotyledons can be gently unfolded. The miniature maple tree will then be discovered nestled between the folded cotyledons.

The unfolding of Norway maple seeds may also be observed on many school grounds in the spring of the year. Frequently, the stalk of the young tree pushes out of the seed coat while the seed is still in the samara lying on the surface of the ground.

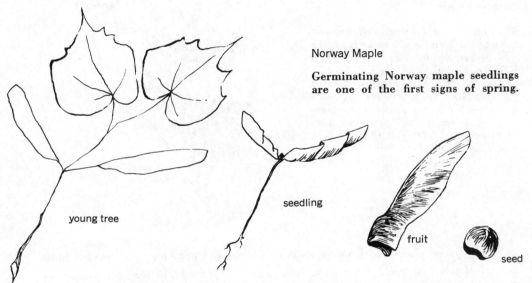

Norway Maple

Germinating Norway maple seedlings are one of the first signs of spring.

young tree

seedling

fruit

seed

Children often call the samara a "Polly nose." Botanists call it a winged fruit. Botanically, the ripened ovary of any flower is a fruit. Fruits contain the seeds of a plant. To a botanist, any plant structure coming from the ovary of a flower and containing seeds is a fruit, regardless of whether it is large and juicy like an apple or watermelon or dry like a bean pod or peanut shell.

The stalk of the Norway maple tree grows downward and develops roots. At the same time, the cotyledons begin to expand and push their way out of the fruit and the seed coat. When they first open, the creases on the cotyledons show where they were folded.

In addition to seed germination, the formation of seeds and fruits by flowers is an excellent topic to study in the spring of the year.

Some plants ripen seeds quickly. Silver maple trees and elms often bloom while snow is still on the ground. By late spring their seeds are ripe and ready to germinate. Dandelions and mustards are also common plants whose seeds ripen quickly.

Most trees take the whole growing season to produce fruits and seeds. The red oaks take two years to ripen their fruits, so two sets of acorns can be found on the branches—tiny ones on the new growth and developing ones on the previous year's twigs. Many of the confiers take two or more years to ripen their seeds, so you may find small undeveloped cones near the ends of branches, green cones ripening seeds father back, and brown cones with open scales that have discharged their seeds on still older wood.

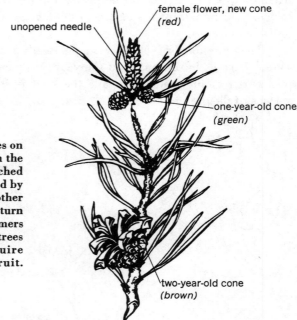

unopened needle

female flower, new cone
(red)

one-year-old cone
(green)

two-year-old cone
(brown)

Scots Pine Twig

The female flowers—small red cones on the ends of branches—emerge from the bud before the needles have reached full size. After the eggs are fertilized by pollen from the male flowers on other parts of the tree, the cones will turn green. Scots pine requires two summers to produce seeds. Look for other trees in your neighborhood that require more than one year to ripen fruit.

Some seed plants, such as dandelions, bloom and produce seeds over a long period of time, so there are always some ripe seeds ready to take advantage of favorable conditions and, conversely, there are always more coming along if conditions are bad and one batch of seeds gets destroyed. On other plants, the seeds may all ripen at one time but remain attached to the plant until something tears them loose, with the result that they get carried in different directions at different times and thereby stand a better chance of falling on good ground.

The various techniques of seed dispersal are always an intriguing topic. Although more seeds are ripe in the fall than at any other time, there is no month in the year when seed dispersal cannot be studied. In winter, seeds travel over the snow on open pods like riders on a toboggan; in spring, they sail on streams of water from melting snow; in all seasons, they get tossed and carried by wind or by animals. They are started on their way by exploding fruits in spring, fall, and summer. The variety of travel techniques reflects the great variety of plants. Many seeds travel by several different techniques.

Plants produce huge numbers of seeds. An idea of potential plant population explosion can be gained by bringing a single plant of almost any weed into the classroom, dividing it into sections and totaling the seeds. Obviously, if all the seeds fell around the plant, they would choke each other to death, but many hazards keep a check and balance on the plant population. Obviously, too, the surplus seeds and fruits are important sources of food for birds and other animals. In fact, seeds offer many possibilities for exploring interrelationships of living things.

RELATED CLASSROOM ACTIVITIES

Possibilities for seed study exist during every month of the year. In the fall, the abundant crop of weed seeds with their great variety of form and many different techniques for dispersal is a topic that catches many teachers' imagination. But seed study can be launched as easily from tree seeds rolling, blowing, and skittering over winter snows, or from the hundreds of germinating seeds in the spring.

If seeds are gathered in the fall, some upper grade children will invariably ask, "Will they grow if we plant them?" If you "try to find out," it is wise to try a great variety of seeds and to keep a duplicate supply for further experiments, for many need a rest period and some must be frozen, while a few, like locust seeds, must have their seed coats cracked. This occurs in the wild when they get tumbled over the ground and blown against rocks.

Just as outdoor seed collections can lead to classroom experiments, so classroom seed studies can lead to outdoor observations.

Any class that has taken a lima bean apart and talked about seeds as baby plants is ready to go outside and watch the "babies" grow in the spring of the year.

Experiences with planting flower or vegetable seeds can be very rewarding. It is wise to use seeds of plants which grow quickly, like dwarf marigolds, radishes or lettuce, so that children can have the joy of seeing their plant produce flowers or of eating their vegetables.

TEACHER PREPARATION

A survey of the area will show what seed possibilities exist. It would be foolish to suggest a field trip to collect wind-dispersed seeds if the only plants available were shrubs with fruits whose seeds are distributed by feeding birds. On the other hand, a field trip can be conducted without knowing every kind of seed that might be discovered. A hand lens will help in the observation and appreciation of all seeds.

Elm, Norway maple, silver maple, and Ailanthus are four of the most common trees around school areas. All of them present excellent opportunities for seed study. A tree guide will help you identify them and learn more about them.

FIELD TRIP POSSIBILITIES

1. Collect as many kinds of seeds as possible and determine how they are dispersed.

2. Collect wind-dispersed seeds. Compare them to human forms of air travel; i.e. propellers, gliders, rockets, wings, parachutes, balloons, helicopters.

3. Collect wind-dispersed seeds. Divide the class in half. With the assistance of a teacher's aide, let one half stay on the school grounds to observe and take notes while the other half launches the seeds from a second-story window on a windy day. Then reverse positions.

4. Examine Norway maple flowers. As the flowers mature, two tiny green structures become visible in the center of the flower. Day by day they increase in length and become the wings of the "Polly nose." Gradually, the petals and their flower parts dry up and the fruits are clearly visible. Often the seed fails to develop and the young "Polly noses" drop to the ground. These can be compared to the ones which remain on a tree producing seeds.

5. In the spring observe germinating seeds. Notice that the cotyledons do not resemble the true leaves. In fact, identification of seedlings is often a wait-and-see operation because only a few cotyledons, like beech and basswood, are distinctive. Many are simply round, like Ailanthus and elm, or long, narrow, and folded, like cherry and maple. After the food in the cotyledons is exhausted and the young plant is started on its way, the cotyledons fall off. Devise a way to discover how long this takes. Is it the same for all seedlings?

6. Silver maples and elms bloom early and ripen seeds before summer. These trees are often gathering places for migrating birds, since they are one of the few sources of fresh seeds at that time of year. It is sometimes possible to observe hundreds of goldfinches or other seed eaters on these trees. Or fruits with the seed torn out may be collected on the ground, which proves that birds or squirrels have been feeding.

7. The whole story of flower to young tree can sometimes be observed with elms. The early blooming flower is wind-pollinated and has no petals, so it often goes unnoticed. The seeds do not need a dormant period and will germinate in the wild or if planted indoors.

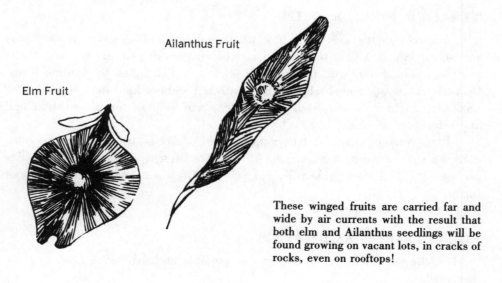

Ailanthus Fruit

Elm Fruit

These winged fruits are carried far and wide by air currents with the result that both elm and Ailanthus seedlings will be found growing on vacant lots, in cracks of rocks, even on rooftops!

8. Observe an Ailanthus tree with clusters of fruit. How big is the cluster in the fall of the year? On a windy day, watch fruits being torn loose and carried away. In what direction do they travel? Are they all torn loose? If some remain, watch on other days. Do they get carried in different directions? How long does it take for the whole cluster to be distributed? How does this increase the chances for some seeds to grow?

9. On a snowy day, observe bird and/or other animal tracks around weeds. Can you find evidence of a relationship between them and seeds?

10. Make a map of seeds of one kind scattered on the school ground. In the spring see how many germinate. Two weeks later, are they all growing? A month later? Does this help explain why plants must produce many seeds?

Grasses

There are between 4,000 and 5,000 species of grass in the world; about 1,500 of these grow in the United States. This means that every area has many kinds of grasses.

Some kinds, such as corn, oats, wheat, rye, barley, rice, and sugar cane are cultivated as important sources of food.

Others, like Kentucky blue grass, red fescue, and timothy are pasture or hay grasses. Still others are grown for lawns or to hold the soil.

Some grasses are classed as weeds, but even weed grasses make and hold the soil, produce oxygen, and frequently serve as food for birds and other wildlife.

Although grasses vary greatly in size, from the 2-centimeter tall species of the tundra to the giant tree-like bamboo of the tropics, they have a great many common characteristics. All grasses have long, slender tapering leaves with parallel veins. The base of these leaves is wrapped around the stalk and is called the sheath; the top part extends from the side of the plant and is called the blade. Grass flowers have no petals or showy parts, but the dangling pollen-producing stamens and feathery pistils are often brightly colored and are very lovely under a hand lens.

A few kinds of grass have the stamen (male flowers) and pistils (female flowers) in different clusters or even on different plants. Thus the tassel of corn is made up of hundreds of pollen-producing flowers. Farther down on the stalk, the ear is made of hundreds of female flowers, each consisting of an ovary and a long slender projection (the corn silk) whose sticky tip will catch the pollen and enable it to travel down to the single egg in the ovary. After the egg is fertilized, a kernel of corn develops in exactly the same way as grass seeds develop on all other grasses. Mowing often prevents lawn grasses from blooming and producing seeds.

The growth pattern of grass stalks is unique. Most plants grow by adding to the tips and ends of branches (terminally), but grass grows at the base of sections which are telescoped inside each other and enclosed by the leaf sheaths. Through the centuries children, as well as many adults, have enjoyed pulling

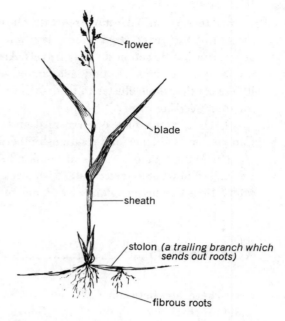

Grass Parts

The seeds produced by grass flowers are the grain crops of the world, providing humans, birds, and other animals with a large part of their food supply. Grass roots are some of the most effective soil holders, fighting erosion on hillsides, sand dunes, and other exposed areas.

these sections out of the grass stalks and nibbling the tender growing bases. These tender tips protected from pollution by the leaf sheaths provide an easily accessible taste experience.

RELATED CLASSROOM ACTIVITIES

A kindergarten class in the Bronx discovered grass flowers when they went outside to look for signs of spring. Their excitement at this discovery was so great that the original purpose of the trip was sidetracked. Children found and picked their own grass stalks, then hurried back to the classroom to examine and admire the dancing colored parts further under their big hand lenses. The story they dictated said:

> Spring is green.
> The grass is green.
> The grass has flowers.
> They are pretty.

Grass flowers and seed spikes are often an experience in beauty. The recording of this beauty may take the form of drawings, blueprints, the projecting or drawing of grass shadows, photographs, stamp-pad prints, mounted specimens between wax paper or clear contact paper, paintings, poetry, or prose.

Experiences with grass flowers may start in the classroom if the teacher or some child brings specimens to school, or they may grow out of a classroom study that involves a planned field trip, or once in a while they may be as spontaneous as the kindergartners' experience.

Often, grass field trips can grow out of social studies units. Topics like cereal foods, the Great Plains, and erosion, all can be enhanced by some actual experiences with this large and interesting group of plants.

TEACHER PREPARATION

In both rural and urban areas the varieties of grass found on a school ground are often surprising. A little time spent examining the lawn and looking for wild grasses or even crop grasses along the edges and against buildings may reveal many teaching possibilities.

Many states have publications about local grasses, their importance and identification. These may usually be obtained by writing to the extension service of the state university or to the state department of agriculture or conservation.

While most grasses are safe for nibbling, a few, like stink grass, velvet grass, and darnel should be avoided. Some others are tasteless. So it is wise to be sure there are enough easily available stalks of the varieties listed and illustrated in trip 7, p. 38, if you plan to include a tasting experience.

FIELD TRIP POSSIBILITIES

1. In the fall of the year when grass has seeds or in early summer when many kinds of grass are blooming, take a grass census. Although grass blades are all similar in shape, some grasses can be told apart by things like the shade of green, the gloss, the hairiness, and the habit of growth of the blades, even if they do not have flowers or seeds. A chart with mounted specimens can be used to display the variety of wild grasses on your grounds.

2. Grasses represent many kinds of seed dispersal. Some, like beard grass, have plumes that help with wind dispersal; others, like sandbur, are sticktights; still others, like the panic grasses, are fine examples of tumbleweeds as they roll

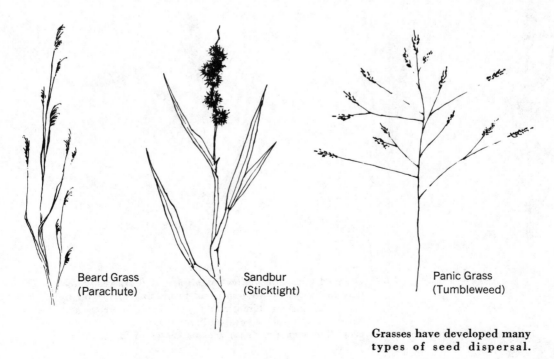

Beard Grass
(Parachute)

Sandbur
(Sticktight)

Panic Grass
(Tumbleweed)

Grasses have developed many types of seed dispersal.

over the ground shedding seeds when their delicate branches bang against the earth. Many others are carried by rain water, stored and forgotten by animals, or scattered by the activities of feeding birds or by people. A class can make a special study of grass seed dispersal or include it in a general study of seeds.

3. The problems of the Great Plains and the dust bowl can be brought into focus by comparing a seeded area with a bare area both during and after rainstorms and windstorms. A grass plant should be dug up and the root examined to fully appreciate its soil-holding qualities.

4. Wild grasses are an important source of bird food. Estimate the amount of food produced on your grounds by counting or weighing the number of seeds on one spike of grass and multiplying it by the total number of spikes. If several species are represented, each should be estimated separately.

5. Examine your school grounds to determine if the area would be improved by planting grass seed. If so, carry out this activity.

6. After a snow, examine the area around clumps of grass for bird footprints, grass chaff, and any other evidence of wildlife feeding.

7. Grass can be experienced with all our senses. Touch: by feeling its coolness on a hot day, rolling the round stem between thumb and finger, comparing the feel underfoot of pavement, bare ground, and springy turf. Sound: who can make a grass-blade whistle? Sight: how many different types of grass can you find? Smell: a few grasses have distinct odors; others smell "green" and fresh if you

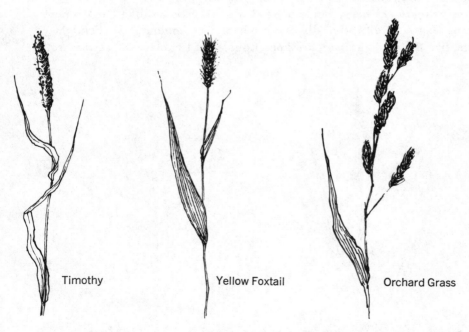

Timothy   Yellow Foxtail   Orchard Grass

Orchard grass and yellow foxtail are wild grasses which may be found holding soil and producing bird food in many waste areas. Timothy also grows wild in urban and rural areas as well as being cultivated as an important hay crop. The stems of these grasses are good for nibbling.

get close to them, particularly in the cool of the morning or after a rain. Taste: three common grasses that are safe for nibbling are timothy, foxtails, and orchard grass.

8. Compare lawn grass to some of the wild grasses. Why don't we use the wild grasses for lawns?

Dandelions

Although a common weed, dandelions add beauty to fields and waste places in the spring.

The toothed leaves with the points slanted toward the ground are unusual. The leaf points and serrations of most plants slant toward the tip. Because of this leaf shape, the French named the plant *dent de lion*, meaning lion's tooth, and from this came the English dandelion.

The flowers, too, are interesting, for each flower head is a cluster of florets like a bride's bouquet. When the green bracts (like the white lace doily of a bride's bouquet) are removed, the individual yellow florets fall apart. At the base of each, a pale green or white seed can be seen developing.

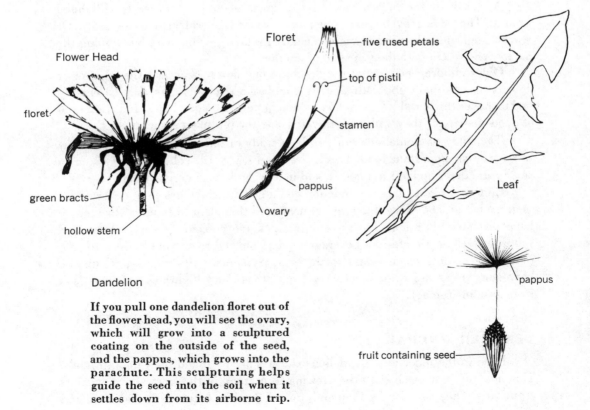

Flower Head

floret

green bracts

hollow stem

Floret

five fused petals

top of pistil

stamen

pappus

ovary

Leaf

pappus

fruit containing seed

Dandelion

If you pull one dandelion floret out of the flower head, you will see the ovary, which will grow into a sculptured coating on the outside of the seed, and the pappus, which grows into the parachute. This sculpturing helps guide the seed into the soil when it settles down from its airborne trip.

Dandelions produce huge numbers of "bouquets." In the spring, flower heads that have died and are ripening seeds, fresh blooms, and an abundance of unopened flower buds will be found on every plant. These buds range in size from large ones ready to pop open to tiny green knobs buried in the center of the plant.

The ripe seeds are interesting not only because of the parachutes which carry them over the countryside on windy days, but also because of the sculpturing and ridges on the seed itself, which, like the thread of a screw, help carry the seed down into the ground.

The hollow flower stem is another unusual characteristic of this plant. It provides a flexibility which enables the flower to bow to adverse conditions rather than break.

The heavy tap root and the fact that the plant is a perennial are other parts of the success story of dandelions.

RELATED CLASSROOM ACTIVITIES

Dandelions lend themselves to an amazing number of activities. The study of dandelions may begin when the teacher brings specimens into the classroom.

Young children may examine the plants to see why they were named after a lion's tooth. They may draw the lion's-teeth leaves and the golden lion's-mane flowers. They love the idea of a bride's bouquet.

A bunch of the hollow stems make great playthings for early childhood groups. The ends may be split open and swished in water to make curls. Or several stems may be used to make a chain, necklace, or bracelet by slipping the narrow end of one into the wide end of another.

Older children may count the florets in one flower head and the number of flowers on one plant and estimate the population explosion. As they go outside and discover the length of dandelion blooming period, plant shape, and technique of seed dispersal, they will see other reasons for the plant's success.

The study of dandelions can also be a study of different cultures, since they were once an important food. Dandelion greens are still collected and eaten in some rural areas and are frequently sold in city markets in early spring.

Children might make a dandelion cookbook from recipes they collect from their grandparents. Since different cultures used this plant differently, they could have an intercultural cookbook with recipes for cooked greens, salads, and soups, as well as a coffeelike beverage made from the roots, and recipes for the roots. An adventuresome class might even try some of its recipes, if suitable harvesting areas are available (away from streets and highways to avoid lead from gasoline fumes).

TEACHER PREPARATION

Even in the inner city, dandelions can usually be found on school grounds. They may be in a patch of grass, growing in a crack in the sidewalk, or against a building. They can also be frequently found in areas between the sidewalk and the curb. Once you locate your specimens, you will be able to decide how to organize your class for observation. Without a trowel, it is almost impossible to obtain dandelion roots.

FIELD TRIP POSSIBILITIES

1. Groups of children may select individual dandelion plants to mark (with something like a numbered popsicle stick) and observe each day. Recorded information can include things like: date of the opening of one flower head, length of its blooming period, time involved for seeds to ripen, date when seed head opened, and number of days before all seeds blew away.

2. On a windy day, watch parachutes take off. How far do seeds travel?

3. Take a census of dandelions on the school grounds. Where are most of them found? Are there places where none grow? Does soil, water, or sun have anything to do with this?

4. Examine dandelions on the lawn before and after the grass is mowed. What characteristics make them successful lawn plants?

5. In protected spots, dandelions sometimes bloom 12 months a year. Check plants on your grounds every month for blossoms.

Making plaster casts of animal tracks in mud is an exciting and engrossing activity. It can contribute to an understanding of fossils, be a craft activity, and, in the interpretation of the track stories, involve oral and written language arts.

Animals

Few things can compete with animals as a source of interest for children. Because of this, many classrooms have pets like hamsters, gerbils, white mice, goldfish, and lizards.

Children can begin to learn about the requirements of life from pets of this type. They also can learn to take responsibility for other living things. However, if children's only contact with animals is restricted to classroom pets and domestic animals, it will be difficult for them to grasp or to appreciate fully the great variety and importance of all living things. In fact, with advertising stressing the evils of all insects and invertebrates, some adults as well as children often forget that life on this planet is possible only because of the activities of a vast variety of interrelated creatures.

Animals range in size from thousands of kinds of microscopic protozoa and worms which live in the soil and water, to birds and mammals of various sizes. Many easily observed animals are invertebrates, animals without a bony interior skeleton.

Invertebrate animals commonly seen on land include: snails, slugs, worms, insects, sowbugs, pillbugs, millipedes, centipedes, and spiders. These small animals far outnumber the better known vertebrate animals both in species and in numbers.

The vertebrates which may be found on land belong to one of four groups or classes. These classes are: mammals, birds, reptiles, and amphibians.

The number and kind of animals on a school ground, as well as the size of any population, will depend on many things. Some primary factors would be: climate of the area, density of the human population in the area, food supply, the availability of a suitable habitat for rearing young, and protection from enemies.

Even the most barren school grounds usually have some hardy animals either as residents or visitors.

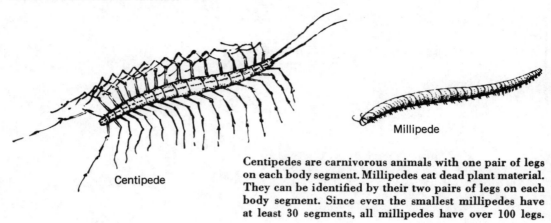

Millipede

Centipede

Centipedes are carnivorous animals with one pair of legs on each body segment. Millipedes eat dead plant material. They can be identified by their two pairs of legs on each body segment. Since even the smallest millipedes have at least 30 segments, all millipedes have over 100 legs.

RELATED CLASSROOM ACTIVITIES

A first-grade class in a New York City school that had only a cement play area had been studying animals. They had taken care of all the common classroom pets. They had made a list of the things their animals needed. The teacher wanted them to relate their learning to the natural world. One winter day, she suggested that they go outside and see if they could find any animals.

The children stood on the snow-covered grounds and looked around. There were no animal tracks, no birds on the one bare Ailanthus tree growing in a pavement crack, no visible animal life of any kind.

This might have been the end of a very successful field trip: lack of shelter and food correlated with lack of animals. However, this building was located next door to a small church. One side of the church had an unpruned privet hedge. In front of the church were two young spruce trees.

As the class approached the church, the chirping of sparrows became audible. The children moved slowly and quietly. They got close enough to see and count 20 birds feeding in the privet. Suddenly, the birds flew from the hedge; some went into the spruce trees, the rest flew to the open belfry.

A few minutes later, excited youngsters were back at their desks enthusiastically drawing the story of the relationship between birds, food, and shelter.

TEACHER PREPARATION

At first thought, planning an animal trip seems difficult. How can you count on the bird or the butterfly that you observe at 8 A.M. being at the same place at 9:30 A.M.? Frequently, you can't. But if you locate a habitat, you can count on things happening. A flower bed in bloom will have some insects coming to feed. It may not be the butterfly you saw in the early morning, but your class may discover many other exciting animals if you turn them loose to do their own looking and make their own discoveries.

Activity around an anthill goes on constantly during daylight hours. A nesting bird will stay around to be observed if the class does not get too close. Life under a log, a board, or a rock is always ready to be discovered.

Sometimes when something really unusual appears, schedules and curriculum are best laid aside. Things like a battle between red and black ants, a mating swarm of flying ants, a cockroach migration, or a flock of cedar waxwings in a hawthorn bush can be introduced with a brief "Let's go look," and can provide a basis for art, science, oral reports, stories, poems, library research, and/or arithmetic.

FIELD TRIP POSSIBILITIES

1. Take a census on your school ground to see how many different kinds of animals you can find.

2. If there is a grassy area, dig up a 0.5-meter cube. Count the animals present. Do a similar survey of a 0.5-meter cube of bare soil.

3. Take a fall, winter, and spring survey in the same area including facts about the lawn, a maple tree, and an evergreen tree.

4. If your school ground has abundant vegetation, take a census of animals living in different habitats like a tree, bush, flower bed, or lawn. Can you find special adaptations that enable different animals to live in the same habitat? Organize your animal list into categories like fliers, climbers, and burrowers.

Vertebrate Animals

A pigeon bobbing its head as it walks busily along, a tree toad croaking loudly in a bush outside the classroom window, a squirrel cutting down a pinecone bird feeder and lugging it away, a red-backed salamander hiding under a board, a chipmunk popping in and out of the cavities in a stone wall, a gull or a turkey buzzard on motionless wings, house sparrows having a family squabble, a mole tunnel on the lawn, a lizard dashing up a tree: All of these are things that have happened and will continue to happen on many school grounds. Some of them, or ones much like them, have happened on yours.

Just which animals and activities can be found on your school grounds is very much a matter of geography and climate, as well as your location in terms of a rural or an urban setting.

In fact, if you were to give an accurate report on the kinds and number of vertebrate animals on your school grounds and describe their activities to an ecologist, he could tell you much about the plants, pollution, soil, location, size, and surrounding area of your school site.

No school ground will have as many vertebrate animals as invertebrates. After all, it takes thousands of insects to keep a mole, shrew, bird, or toad alive. Plant eaters also need abundant food supplies.

Red-backed Salamander
7 to 13 cm long

American Chameleon (Anolis)
10 to 20 cm long

Although salamanders are amphibians, and lizards, like the American chameleon, are reptiles, they are often confused because they have a similar body shape. They have many distinct differences, however, which children can quickly learn to recognize.

SALAMANDER	LIZARD
moist skin	dry skin
smooth skin	covered with scales
no claws	claws
ridges on outside of body	no ridges on outside of body
prefers darkness	likes bright sunlight
hides under rocks and logs	climbs trees and fences

All animals need shelter. Some school areas do not have any suitable areas for hiding, building homes, rearing families, or resting.

Many vertebrates go about their business at night or in the very early morning. Even those that are active during the daytime, like squirrels, snakes, and birds, tend to avoid people and either hide or stop their activities when noisy, active people approach.

Successful observation of vertebrates will depend on three things: training a group to approach and observe with minimum motion and noise (stalking); observing common animals like squirrels, house sparrows, and pigeons, which have adjusted to humans and often go about their business undisturbed; and learning to read signs and tracks. Because of the almost universal opportunity for animal track and bird study on the school ground, these two subjects each comprise a separate topic in this section.

RELATED CLASSROOM ACTIVITIES

A study of interrelationships and food pyramids could trigger a field trip to compare the number of vertebrates in an area to the number of plants and invertebrates upon which they depend.

A study of Indians or pioneers who depended on hunting skills could serve as a challenge to the children to try their skills at observation, silence, body coordination, and the slow movements necessary for learning about wild animals.

An interesting exchange of data could develop between two schools in similar situations in different parts of the country or two schools in the same part of the country in different situations. For example, a third-grade class in a rural school in Kentucky could write to a third-grade class in a rural area of New Hampshire and exchange information. This would be particularly meaningful if both classes stressed their observations of animal behavior.

TEACHER PREPARATION

As always, a familiarity with the school ground and its potential is important.

For animal studies, some teachers also need a psychological preparation. Often adults have been conditioned to be afraid of the natural world. This is unfortunate and difficult (but not impossible) to change. It is important not to pass these feelings on to youngsters. The teacher who is terrified of small animals does not have to handle them, but she should be prepared to say, "Let's put the animal in a jar so we can keep it safe while we all look at it." Even if the teacher is at ease with the animal, it is often better for it not to be passed from one hot little hand to the next. This is particularly true of moist-skinned amphibians.

Reptiles are rarely found on school grounds since they tend to avoid people; but if you really want to keep and observe a snake or a lizard even for a brief time, putting it in a glass container minimizes the chance of escape and permits maximum observation by all.

FIELD TRIP POSSIBILITIES

1. Squirrels frequently have feeding lookouts on a rock, a place on a wall, or a horizontal limb where they can survey the area for danger while they eat. These are marked by piles of nutshells and other food remains. Can you locate one on your site? What did the squirrels eat? Where did they get the food?

2. Conifer seeds are a favorite squirrel food. Squirrels obtain them by peeling off the scales of pine, spruce, and other cones and eating the two seeds

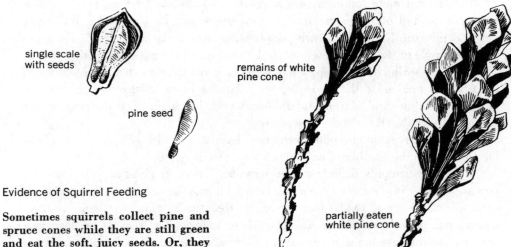

single scale
with seeds

pine seed

remains of white
pine cone

partially eaten
white pine cone

Evidence of Squirrel Feeding

Sometimes squirrels collect pine and spruce cones while they are still green and eat the soft, juicy seeds. Or, they may store the cones for winter use.

on each scale. Bare or half-bare cone shafts can often be found under evergreen trees or on feeding lookouts.

3. Examine the ground under trees for short twigs that tell of squirrel activity. Sometimes when branches are too light to support their weight, squirrels go out as far as they can, then bite off the tips to get at cones, like hemlock, or at tree buds fat with small leaves or flowers for the coming year. Then they hurry to the ground to feed.

4. Try to discover a mouse feeding area in a protected place on a wall, under a rock, or in a weed patch. Squirrels crack nuts; mice gnaw round holes in nuts. Discovering cherry pits or nuts with round holes gnawed in them reveals the nighttime activity of mice.

Mouse Activity

Squirrel Activity

Mice nibble holes in the sides of nuts and fruit pits; squirrels use their teeth as a nutcracker.

5. Go outside to locate squirrel nests in trees. They are made of leaves and twigs. In winter, squirrels prefer to live in holes in trees, but if none are available, they live in nests with tight walls. Summer nests are loosely made and larger ("air-conditioned").

6. Follow a mole tunnel across a grassy area. Moles feed on beetle grubs, earthworms, and other ground-dwelling invertebrates. They help air and water get into the soil, but they disturb people who want their lawns level. They are a fine example of the complicated interrelationships of the natural world.

7. In the spring after the snow melts, many lawns have a network of pathways, the trails of field mice that moved just under the snow all through the winter. Have every child find a trail and follow it with his hand. Try to imagine the mouse activity that these trails represent.

8. Do a survey of possible vertebrate habitats. Could you do some planting to increase the number of animals on your school ground?

9. If your grounds include a flower garden, lawn, or shrubbery border and you are located in a rural or suburban area, you may be able to find amphibians such as salamanders, toads, tree frogs, or tree toads. Observe them in their natural habitat. If you decide to bring them indoors for a visit, be sure you have an appropriate home prepared for them and can provide live food. These animals need a moist environment so a terrarium with plants in which water

condenses on the top every day is ideal. But don't put toads into a highly prized planting! Their digging habits are fascinating to observe, but they do uproot the plants. Outdoors this problem does not exist, since there are plently of plant-free hiding places.

10. Can you find evidence of domestic animals visiting your school ground? Why do they come?

Birds

Birds are the best known of all the wild vertebrate animals. There are many reasons for this. Birds are unmistakable: no other animals have feathers.

Furthermore, birds are animals of the air. Most of them fly. The nonfliers run over the ground. Most of them build their nests in trees, bushes, or on the ground. The few, like bank swallows and kingfishers, who nest at the end of tunnels are conspicuous fliers whose presence can never be missed.

Except for owls, birds are daytime creatures. Even owls announce their presence with loud hoots. The songs, the crowing, the warning calls of other birds all help to make people aware of them. Some birds are brightly colored; and even the drab ones are generally graceful and add beauty to the landscape. Furthermore, birds are found everywhere from the frozen wastes of Antarctica to the tropics, and from the North American tundra to the temperate zone. They are found in field and forest, city and suburb.

It would be hard to find someone who has never seen a bird, but it is easy to find people who know relatively little about the wonderful variety and habits that this group of animals represents, or the complicated interrelationships of which they are a part.

All too often bird study consists of trying to see how many kinds of birds can be seen in an area, instead of trying to learn as much as possible about the life style of one or two birds.

Bird study does not require rare birds. In fact, some of the common birds like house sparrows, starlings, jays, robins, and pigeons are ideal for study since all live near people and usually can be observed going about their activities.

RELATED CLASSROOM ACTIVITIES

Ornithologists (people who study birds) often hide in specially constructed shelters called blinds to observe, photograph, and take notes on the activities of specific birds. Some classrooms and school hallways with windows can serve as blinds. Children can stand by the windows to draw and take notes on the behavior of the birds outside. If the windows overlook nests, feeding trays, or other places where birds congregate, the children may be able to observe more from this vantage point than they could outdoors where their presence would be disturbing.

Birds may be attracted to your area by the children planting shrubbery, like holly, red cedar, hawthorn, and Russian olive, which provides both food and shelter, or by putting up feeding stations. Interesting feeders for small birds can be made by stuffing suet or peanut butter between the scales of pine cones and hanging them from trees or from wires. These small swinging feeders delight chickadees and nuthatches but will not be touched by house sparrows and pigeons. Your class may, however, find themselves pitting their wits and engineering ability against the agility and keen mind of a greedy squirrel.

Pine Cone Bird Feeder

Feeding stations—such as this pine-cone feeder for small birds—can be made by young children. They are composed of suet or peanut butter stuffed between the scales of pine cones and hung from a tree or wire.

A bird bath filled with clean water is a fine way to attract birds. This can be made from a garbage can lid or it can be constructed by digging a shallow basin ranging in depth from a half inch to three inches and lining it with concrete.

Dust baths are also enjoyed by birds. These are made by digging up a patch of soil and pulverizing all the lumps in an area that is fairly dry.

In the spring, nesting materials can be put on a feeder which is no longer needed, or in any accessible spot that can be viewed from the classroom. Straw, packing materials, 15- to 20-centimeter-long pieces of string or yarn, feathers, bits of cotton, and hair are all good nesting materials. If different birds come for the materials, do they take different materials?

Birds nests are really nurseries for rearing young. Once a brood has been raised, the nest is abandoned. Birds that raise two or three broods a year build a new nest for each brood. Except for house sparrows, which select a nest site and keep it for their lifetime, most birds have no interest in abandoned nests. By moving to a new site and using clean bedding, mites, lice, and disease are left behind. Even house sparrows build new nests for each brood. Watch a female throw old material out of her nest after a brood has been raised. Watch her supervise the materials the male brings, discarding what she considers inappropriate. Try to identify the materials that are being used for nest construction. Are some of them modern materials? Might building materials in a city be different from those of a rural area? What information might you exchange with children in a different environment about bird observation?

TEACHER PREPARATION

If you observe your school ground for a short time both before school begins and from the window during school hours, you will discover some areas where birds habitually congregate. These will be good spots to take your class for making observations. If the children move slowly and quietly, they can frequently get very close to the birds. Even if some birds fly away, they often return if the class waits patiently.

FIELD TRIP POSSIBILITIES

1. Go outside and watch house sparrows and pigeons as they travel over the ground. Let children pretend they are sparrows or pigeons. Devise games based on impersonation and identification. Observe both types of birds in flight. How do sparrows glide? Pigeons? Let the children use their arms to demonstrate flight patterns. Compare the behavior of other common birds.

2. Make a map of the school ground and indicate the places birds commonly congregate. Correlate this with the need for food, water, and shelter.

3. Examine tree trunks for woodpecker and sapsucker drilling. Woodpeckers' holes are scattered; sapsuckers' holes occur in horizontal lines. Each sapsucker hole is a little sap well from which he drinks; each woodpecker hole represents one insect taken from beneath the bark.

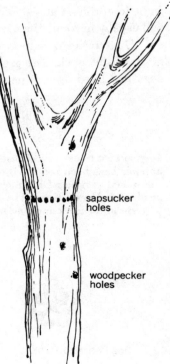

Woodpeckers dig holes wherever they hear an insect eating under the bark; sapsuckers drill a row of sap fountains to which they return repeatedly to drink.

sapsucker holes

woodpecker holes

4. If there is a bird nest on the school ground, try to devise some technique whereby committees of children can observe and count the number of feeding trips that the parent birds make in a school day. This will involve putting the bird watchers in a place where they can see but where they will not distract the birds.

5. If flocks of birds stop on or near the school ground during their spring or fall migration, take advantage of this unusual occurrence and go out and watch.

Animal Tracks

Many animals are active at night. Others that are active in the daytime are shy and hide when people are around. To learn about animal activities, hunters and naturalists through the ages have learned to read the signs that animals left behind. These include tracks, droppings, feeding places, nests, and burrows.

The most informative of these, in terms of animal activities, are usually tracks. In identifying tracks, the shape of the paw, the presence or absence of nails on the paw, the size of the print, the comparison of front and hind prints, the distance between front and hind prints and between one set of prints and the next are all important. Frequently, tracks are not perfect, and even perfect tracks lose their detail as time passes. But if a combination of characteristics is used, even children can identify many animals by their tracks.

Cats and foxes both walk in a straight line and put their hind feet in their front footprints, but size of track, presence of nail prints in fox tracks, and area will usually make distinguishing one from the other easy.

Dogs make a pattern similar to cats but their hind feet seldom squarely hit the front track. The great American naturalist Dr. E. L. Palmer used to say dogs had lived around people too long and had gotten sloppy.

Dog and cat tracks reveal that these animals have been recent visitors to the school grounds. Dog tracks can be identified by the size and irregular patterns of the track and claw marks.

Dog Print and Track   Cat Print and Track

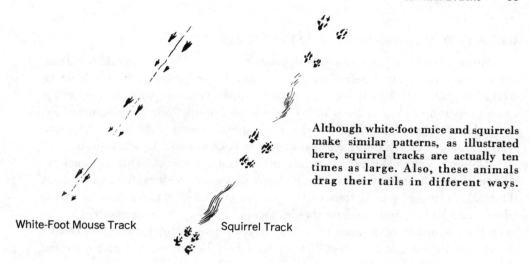

Although white-foot mice and squirrels make similar patterns, as illustrated here, squirrel tracks are actually ten times as large. Also, these animals drag their tails in different ways.

White-Foot Mouse Track   Squirrel Track

Squirrels and white-foot mice both hop by putting their small front feet down beside each other and swinging their larger hind feet in front of them. However, no one would confuse the size of a squirrel print with a mouse print.

Rabbits and field mice put one forefoot in front of the other, then swing their hind feet forward. Often when snow conditions are right, mice leave a tail print.

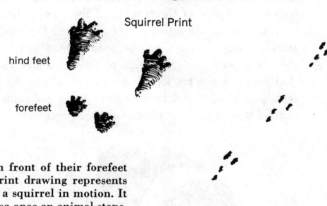

Squirrel Print

hind feet

forefeet

Rabbit Track

Animals swing their hind feet in front of their forefeet when they hop. The Squirrel Print drawing represents one set of tracks from a series of a squirrel in motion. It could never be a single set because once an animal stops, his forefoot marks would precede his hind ones. Rabbits, field mice, and other land dwellers put one forefoot in front of the other in hopping. Squirrels, deer mice and other tree dwellers put their forefeet down together.

Identification of the animal is only one part of the animal track story. More important is finding answers to questions like: "Where was the animal going?", "What was he doing?", "Did he travel across open space or around edges?", and "Did he find food?" Sometimes a drama involving several animals can be discovered, and excitement runs high.

Children love to feel they have a kinship with hunters of old, detectives, or Indians; but tracking is more than just fun. It is a good way to learn careful observation, to begin to understand the interrelatedness of all animal and plant life, to amass evidence, and to draw some conclusions.

RELATED CLASSROOM ACTIVITIES

Since a tracking trip frequently depends on two highly unpredictable things, the weather and animal behavior, a class may be prepared and then have to wait a long time. Or the trip can be a surprise to the children, with the only classroom preparation an invitation to go outside and learn, "what the squirrel did early this morning," or "how we can read animal tracks." After the trip, the stories read from fresh tracks may be written in essays or told in oral reports.

Children can make potato prints of the animal tracks. This is done by slicing a potato in half and drawing the track on the cut surface, then cutting out the track drawing from the rest of the potato. (Care, of course, must be taken when using knives, and children should always cut away from themselves.) The print thus made can be pressed on an inked stamp pad. A total track pattern like the grouped front and hind feet of a squirrel may be cut on one potato or the print of one foot like a dog or cat paw print can be made. Potato prints can be used to record the animal tracks which were observed or the children can compose new stories using the same animals or adding others. Using the potato prints, one group can illustrate a story for another group to interpret. Large murals on newspaper or brown paper may tell many animal stories, since many things occur simultaneously in nature. Children can begin to see the relationship between the size of the track and the size of the animal, the distance between tracks, and the direction of toe prints. Other easily cut materials such as cardboard or large erasers may be used instead of potatoes.

If plaster of Paris molds of animal tracks have been made outdoors, exact duplicates of the tracks can be made indoors by completely covering the mold with a coat of Vaseline or other greasy substance, putting a cardboard collar around it, and pouring plaster of Paris over it.

potato

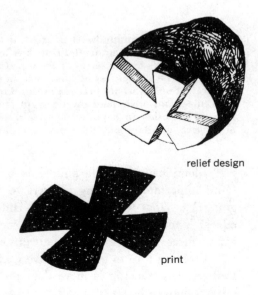

relief design

print

Potato Print

Making prints of animal tracks is an activity which can be done any time of year. Potatoes and other root crops can be used for these prints.

TEACHER PREPARATION

Alertness and flexibility are the best preparation for the study of animal tracks. In most school situations, tracks are not available just any time.

Furthermore, tracks do not last. Snow melts, more snow falls, or it rains. In areas of high population density (and most school grounds are), tracks get buried under other tracks. So preparing for an animal track lesson usually involves waiting, watching, and grabbing the opportunity.

FIELD TRIP POSSIBILITIES

1. Go outside in a light snow. Have a small child and a larger one go around the corner of the school building (so they are not seen by the class), walk across the snow, and return. Let the class identify the tracks of each. Have two children go around the corner of the building and instruct one to run and the other to walk. Can the children tell which tracks represent speed? What else can they learn from observing their own tracks?

2. Follow a squirrel's tracks in the snow. Where did the squirrel come from? Where did it go? Did it return to its starting point? Can you draw any conclusions? (Does it have more than one winter home? Is it still out hunting? Might it have gone home overhead?)

3. Look for bird tracks along a fence, in a weed patch, and along bushes. What conclusions can you draw about the size of the bird? Its method of traveling on the ground? Can you find a relationship between tracks and food? Tracks and shelter? Can you find tail and wing marks where birds landed?

4. Take a census of the number of dogs that crossed an area by the size of their prints. If several dogs were in the area, were they there at the same time or different times? Did they pay any attention to each other?

5. After a snow, do a study of people tracks on a sidewalk. Can you tell size, age, speed, direction of travel, or activity?

6. Go outside after a light snowfall and let the children use their hands and feet to make common animal tracks. Walk like a dog, cat, starling, pigeon, and squirrel; hop like a squirrel, rabbit, and sparrow.

7. Using plaster of Paris, cast a track. For outdoor use, put one inch of water in the bottom of a container; then add two inches of snow. Stir. Pour plaster of Paris so it makes a peak that extends an inch above the water. Stir. This mixture should be like whipped cream. If it isn't, add more plaster. Put a cardboard collar around the track. Pour the plaster so it gently flows into the print. Let it harden (this will take an hour or more).

8. Discover track stories in mud or fine dust.

9. If there are enough tracks of one animal to compare distance between sets of prints, can you draw any conclusions about speed of travel? Can you discover any reasons for the animal to be moving slowly? Quickly?

10. Can you find a track story that involves two different animals?

Insects and Other Arthropods

Many people would be amazed if they had any idea of the number of animals which surround us. Most of the animals of the world are small; some are microscopic. Some are active only at night; still others carry on their activities in the soil, inside a plant, under a rock, or in some other concealed area. All of them contribute to the complicated story of the interdependence of living things.

Zoologists have put animals with similar characteristics into groups. The large groups or divisions are called phyla (singular: phylum). All animals with an exoskeleton (external skeleton), jointed legs, and segmented bodies belong to the phylum Arthropoda (Greek, meaning jointed foot). There are more named arthropods in the world than any other group of animals, but there are still huge numbers that have never been studied or given a name.

The big phylum Arthropoda is divided into six classes; five of these classes consist of animals that live on land. They are: insects, arachnids, crustaceans, millipedes, and centipedes. Telling the classes apart is a matter of counting.

Adult Insect

Body Parts

Appendages

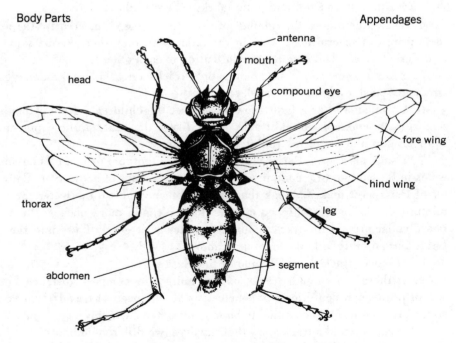

antenna

mouth

head

compound eye

fore wing

hind wing

thorax

leg

abdomen

segment

Insects can be distinguished from all other animal groups by their segmented bodies, six legs, and three body parts.

Spider

Daddy-Longlegs

Daddy-longlegs, one type of arachnid, has long slender legs; all its body parts are fused into one. Spiders, another common arachnid, come in many shapes and sizes but all have two main body parts—the cephalothorax with the eyes, mouth, and legs, and the abdomen, with the silk-making spinnerets.

Many mites are microscopic; all mites are small. Young ones have two or four legs; mature ones develop eight.

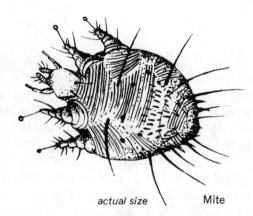

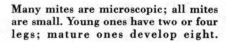

actual size \ Mite

Insects have three body parts: head, thorax, and abdomen. If they have legs (and all insects do at some stage in their development), they are six in number and are attached to the thorax. If they have wings, they, too, are attached to the thorax. Wings always indicate an insect and they also indicate adulthood. No insect with wings will ever grow any larger.

Arachnids have the head and thorax fused into one unit called the cephalothorax (Greek, *cephalo* meaning head). Attached to this are eight walking legs. Spiders, daddy-longlegs or harvestmen, mites, and scorpions are all arachnids. In front of the eight walking legs is a pair of appendages which may be small and inconspicuous as in daddy-longlegs and mites, or large and very conspicuous as in scorpions, or of varying sizes as in different kinds of spiders. Though they sometimes look like extra legs, these appendages are used for grasping food, mating, and feeling. They are called pedipalps (Latin, meaning foot-feeler).

Crustaceans do their feeling with two pair of antennae. They breathe with gills. One would hardly expect to find animals breathing with gills on land; however, two kinds of crustaceans, the pillbug and the sowbug, do breathe this way. These two small animals are found in damp spots. Pillbugs roll up in a ball when touched. Sowbugs are unable to do this. Like their relatives, the crayfish and lobster, pillbugs and sowbugs carry their eggs glued to the underside of their abdomens.

Pillbug

startled pillbug

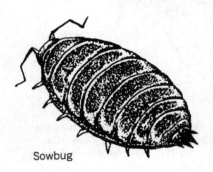

Sowbug

Sowbugs and pillbugs are our only land crustaceans. Commonly seen on land, they look alike, except that pillbugs have the ability to roll up into a "pill" when disturbed. Both animals must always be moist because their breathing apparatus is similar to the gills of their aquatic relatives, crayfish, crabs, and lobsters.

The last two classes of arthropods are the leggy members of the family. Centipedes (meaning hundred feet) have one pair of legs on every body segment. The house centipede has 15 segments; other centipedes have more body segments and, therefore, more legs. Centipedes have long legs which break off easily if an enemy grabs them. They move fast and live on the small animals which they eat.

Millipedes (meaning thousand legs) have two pairs of legs on every abdominal segment and one pair on each segment of the thorax. Adult millipedes have at least 30 leg-bearing segments; young millipedes have less. Millipedes move slowly with a rhythmical pattern which seems to flow along their short, beautifully coordinated legs. When they are disturbed, they curl up with their underside protected. They eat only dead matter.

How many of these classes can you find represented on a school ground? Usually all, except on a concrete-asphalt desert where life will be limited to those animals which can fly or crawl onto the area and feed on human discards.

Are arthropods dangerous? Rarely. Black widow spiders, tarantulas, scorpions, and tropical centipedes can inflict a painful and serious injury on anyone who picks them up, but as long as they are undisturbed, they do not bite or sting. Even if they are disturbed, they will use escape as their first line of defense if at all possible. In addition, some bees and wasps, including the wingless wasps called velvet ants, inflict a painful sting when disturbed or handled.

Recently, a great deal of publicity has been given to the extremely small number of people who have a serious allergic reaction to bee stings, with the result that many people are unduly frightened and panic in the presence of a bee or wasp, or feel that they should all be eliminated. Again, stinging is a defense and is used only by animals that feel they are being attacked. Honey bees die after stinging, so the individual is sacrificed for the group. As long as no one swats at bees gathering nectar and pollen, or otherwise threatens them, the bees can safely be observed going about their business.

pedipalp

Scorpion

Like all arachnids, scorpions have eight legs, but their huge grasping pedipalps look like another pair of legs. Their stinging tail makes it easy to distinguish them from all their relatives.

An indiscriminate fear and dislike of insects has been nurtured by insecticide advertisements, which imply that getting rid of all insects would be a great benefit to humankind. This, of course, is untrue. Bees play an extremely important role in crop production and the complicated interrelationships between insects and soil, plants, other animals, and the total web of life are just today vaguely being sensed by scientists and ecologically minded persons.

The wise person will know and respect the hazards of an area, then relax and enjoy the great numbers of interesting, exciting creatures that contribute to the web of life on this planet.

RELATED CLASSROOM ACTIVITIES

Insects are some of the best classroom pets. They are small, easy to maintain, cheap, clean, and generally odorless. Furthermore, things happen.

A caterpillar that spins a cocoon or makes a chrysalis and emerges as a moth or butterfly in the classroom provides several exciting events. A praying mantis which constructs an egg mass before a fascinated class, a cricket that molts its skin, a bagworm that makes a silken case decorated with plant materials, all provide constant opportunities for learning.

An insect or arthropod zoo can result from a collecting trip to the school grounds, or a short field trip can be introduced by observing classroom animals.

The story of complete metamorphosis can be learned firsthand from mealworms, the larvae of the darkling bettle. These larvae are sold in pet stores as food for frogs, lizards, and other carnivores. If they are placed in a container with some grain product like bran or oatmeal, they grow rapidly, pupate, emerge as beetles, mate, and lay eggs which hatch and continue the cycle. A slice of apple or other vegetable material should be put on the surface of the meal to provide moisture.

Because food for them is easily available at any season, they can be grown any time or used for experiments over a long time. Insects that feed on plants that lose their leaves in winter or on other insects should be permitted to hibernate. This can best be done by putting pupae, egg masses, and other winter forms in a container where birds and other hungry animals can't reach them and storing them until spring in a cold place such as in a garage, on a windowsill, or a fire escape, or in a refrigerator.

Housing for insects can be made from any clear plastic or glass containers. Holes for air circulation can be made in the plastic by heating a knife and pressing it against the plastic. Holes for the glass containers can be punched in the lid with a hammer and nail. If the holes are small, they need not be covered.

For leaf-eating animals, the only furnishing needs to be a small vial or bottle for water, covered with a piece of cardboard through which twigs with leaves can be stuck.

Many invertebrates will do well in a terrarium. Plants which provide food for insects or other invertebrates may be part of the terrarium or they may be added as needed in a cardboard-covered vial. (The cover keeps the insects from falling or crawling into the water and drowning.)

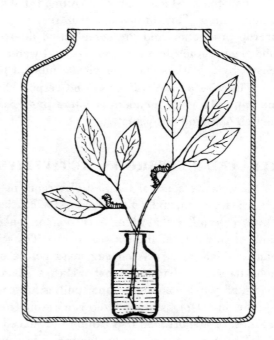

Insect Home

Placing a cardboard over the top of a small vial or jar and sticking the plant stems through it prevents the insects from wandering or falling into the water and drowning.

A terrarium has the advantage of a built-in oxygen and water cycle. Animals in a terrarium do not suffer from desiccation. Furthermore, the total setting is relatively natural. Fecal material falls to the ground and becomes part of the soil cycle. Soil adds to the comfort and general well-being of some animals like those that live in dead wood.

The only disadvantage of a terrarium is the fact that it sometimes reduces visibility, making observation difficult.

TEACHER PREPARATION

Walk around the school ground and look for possible habitats: a lawn, a flower bed, a flowering shrub, a hedge, a tree, a rock on the surface of the soil, a board resting on the ground, a dead log, an anthill in a crack in the pavement. The fall and winter cocoons of moths are often found under the trim of buildings. Decide how you are going to organize your class and what your objectives will be.

FIELD TRIP POSSIBILITIES

1. Go outside and observe the activity around an anthill. Are ants carrying food? What kind? Do they pay attention to each other? Sometimes when an ant colony is disturbed, the eggs, larvae, and pupae can be seen. Watch nurse ants pick up mouthfuls of tiny white eggs, slender glistening ant grubs, or ant pupae and carry them to safety. Some ants pupate in a papery cocoon. Others have a naked pupa which looks like a white, inactive ant with all parts formed including legs.

eggs

larvae (grubs)

worker ant

pupae in cocoons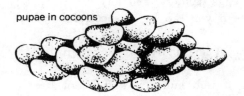

An ant home always contains one queen ant, many worker ants, clusters of tiny eggs, glistening white larvae, and the pupae. In a few species of large ants, the pupae are naked with legs and body parts easily identifiable, but most ant pupae are in papery cocoons which are sold for pet food and often incorrectly labelled "Ant Eggs."

2. Make a study of ant colonies. How many kinds of ants can you find? Give them descriptive names like mini-reds, two-tones, or pyramid builders.

3. If the trees on your grounds have bagworms, do a population study. Bagworms pupate in late summer. The males develop wings, emerge from the cocoons, fly to female cocoons, and mate. The females never leave their cocoons. After mating, they lay eggs in the cocoon, then shrivel up and die. The male cocoons will be empty, with the empty pupa case protruding from the bottom if the wind has not broken it. The female cocoons will be filled with eggs. Count the eggs in one cocoon. Multiply it by the number of female cocoons (half the

Bagworm

There are 20 species of bag-worms in the United States ranging in size from 0.5 centimeters to several centimeters long when mature.

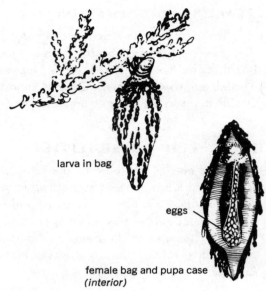

larva in bag

eggs

female bag and pupa case
(*interior*)

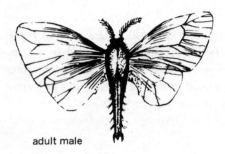

adult male

total). Watch for checks on the population explosion. (Bagworms are extremely interesting insects to rear indoors. They will feed on all tree leaves including evergreens.)

4. Do a survey of arthropods on the school ground.

5. In winter observe a brown creeper, a nuthatch, or a chickadee feeding on a tree. These birds are eating insects, insect eggs, spiders, and any other live animals they can find. Using a hand lens, see if you can be as successful hunters as they are.

6. Make a study of the creatures under a piece of board. Be sure to return the board afterward so as to preserve the habitat.

7. Make a study of spider webs. Each kind of web indicates a different kind of spider. Orb webs may be collected by spraying them with hair spray or the kind of fixative used on charcoal, then putting a piece of black paper against them.

8. Make a study of the insects on a flowering tree or in a flower bed. How many different kinds can you count? What are they eating? How do they eat? Compare bee activity to that of other insects.

9. If you can find a plant with aphids (plant lice), notice the variety in size. Aphids give birth to living young. If you are fortunate, you may see this happen. You may also see ants moving up and down the plants. Ants collect a sweet substance which the aphids produce. If the aphids are threatened by lady bird beetles (ladybugs) or some other aphid enemy, the ants often carry them to safety.

10. If there is an oak tree on your school ground, see how many galls you can find. Galls are plant growths that enclose a developing insect. They are found on many plants but the largest variety are found on oaks.

11. Look for leaf miners, insect larvae that live between the top and bottom layers of leaves. Notice their adaptations to life in a flat area. What are the advantages of living inside a leaf?

12. In the spring, go outdoors and watch caterpillars making silk. Collect caterpillar silk and spider silk; compare them. Compare the way the two animals make and use the silk.

13. Watch a mud dauber wasp build a nest. (Unless you pick them up, these wasps never sting.) See if you can locate the source of mud. How does the wasp dig up the mud? Carry it? Apply it to the nest?

14. Take a winter field trip and see if you can find any hibernating arthropods. Look under windowsills, in bark cracks, in sheltered spots, and under leaves.

15. Put out pieces of food where dogs, cats, and birds will not disturb them. How long does it take insects to discover them? To clean them up? Does this help demonstrate the importance of the earth's cleanup crew? In urban areas, does it help dramatize the slogan "starve them out" in connection with roaches?

16. If the school has a tape recorder, use it to record a singing cricket, grasshopper, cicada, or katydid.

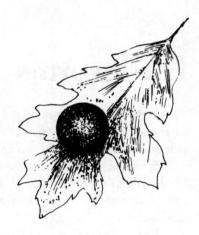

Oak Gall

The oak apple gall is a common and conspicuous gall which is produced by the oak leaf in response to a chemical deposited by the gall insect.

Leaf Mines

Some leaf mines are winding, pale-green trails on leaves; others are round or trumpet-shaped. All house (or housed) a tiny insect larva with a body so flat that it could feed and live between the top and bottom layers of the leaf.

17. Take a trip around your school ground and observe biological controls like ladybird beetles eating aphids, parasitic wasps on sphinx caterpillars, and spiders and praying mantises catching their prey.

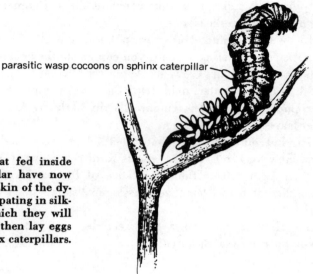

parasitic wasp cocoons on sphinx caterpillar

The wasp larvae that fed inside this sphinx caterpillar have now broken through the skin of the dying host. They are pupating in silken cocoons from which they will emerge to mate and then lay eggs in other young sphinx caterpillars.

Earthworms

Earthworms are one of the easiest invertebrate animals to collect. They can be found almost any place where there is moist earth, even on lawns in front of high-rise apartment houses and on other plots of city soil.

While earthworms are nocturnal animals, they can be located during the day by the piles of castings around their holes, and also by their presence above ground during and after a rain. In cities, they are frequently found stranded on sidewalks.

The earthworm is well adapted to moving through the soil. Its streamlined body offers no resistance; the two pairs of bristles on the underside of each body segment can be extended and retracted. They are important both in helping the worm move and in resisting any effort to pull it from its burrow.

It has no eyes, but the skin cells on the upper surface of both ends of its body are light-sensitive. Its mouth is on the underside just behind the front segment. This mouth is a simple opening through which food is drawn by the muscular pharynx at the end of the esophagus which expands and contracts. It operates very much like a medicine dropper.

Adult worms have a thickening called a clitellum which extends from the thirty-first to the thirty-seventh body segment and produces mucus. This mucus helps keep the body moist, is used to hold two worms together in mating, and forms the shell of the egg cocoon.

Cutting an earthworm in two never makes two worms. If an earthworm's tail is cut off, it grows a new one; but if it is cut anywhere between its head and clitellum, it dies.

Earthworms eat their way through compacted soil. Inside their bodies, the soil is ground fine in the gizzard and organic material is removed by the digestive tract. The pulverized soil is excreted in the form of pellets called castings. In working over the soil in this way, earthworms play an important role in soil enrichment. They also grab organic material like leaves and dead insects and pull them down into their holes. Frequently, parts of these get stuck in the hole entrance and serve as organic fertilizer. Often, seeds get planted in this way. The largest number of earthworms will be found where there is an abundant source of plant material. A forgotten pile of leaves in a fence or by the corner of a building frequently contains many earthworms ranging in size from tiny, pale, almost transparent babies, to young and mature specimens. Baby earthworms differ from adults in size, intensity of color, and number of body segments, as well as the lack of a clitellum.

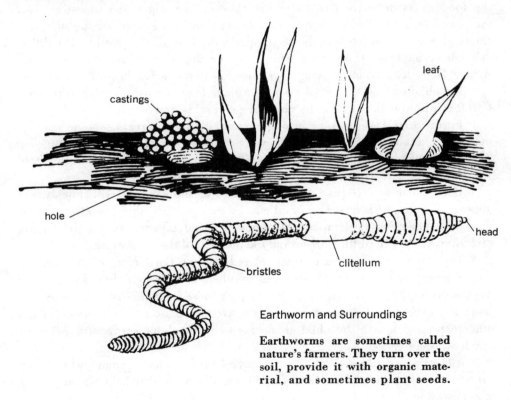

Earthworm and Surroundings

Earthworms are sometimes called nature's farmers. They turn over the soil, provide it with organic material, and sometimes plant seeds.

RELATED CLASSROOM ACTIVITIES

The study of earthworms can be part of a study of animals, adaptation, soil, or the interrelationship of living things. A lesson could grow out of a report on Darwin's *Earthworms,* or a class's reaction to the dozens of worms on the sidewalk on a rainy day. Earthworms could be brought into the classroom before a field trip, or they could be collected during a field trip. Either way, a series of lessons on earthworm behavior can be an interesting experience.

How does the earthworm use its muscles? Which is the head end? As the earthworm moves forward, what does it do? If it followed this behavior pattern on soft earth, what would happen? On concrete? Have the children put earth at one end of a pan. Does the earthworm see the earth at the end of the pan and move toward it? Encourage the children to test the worm several times and check their conclusions with the other groups. Do all the worms behave the same way? If the children place the worm on soft earth, what does it do? Can they explain why dead earthworms are found on sidewalks and other hard surfaces after a rainstorm?

Have the children observe an earthworm for a while and think of words to describe its size and shape. Do they need different words at different times? Why?

The pink color is from hemoglobin in the earthworm's blood. The black is the food in its digestive tract. Can the children see where the digestive tract starts and ends? Is there an equal amount of food in all parts of the digestive tract? Can the digestive tract be seen equally well in all the worms? Do differences have anything to do with size or age or the place where the worm was living? Does the class have enough evidence to come to a conclusion?

Let children make a list of other questions they can ask about earthworms. Can they discover the answers by observing several worms?

Earthworms may be kept in a terrarium, or an observation home may be made from a tall, straight-sided, narrow jar of the type that olives often come in. The jar should be filled about three-quarters full of moist earth, and enclosed in a sleeve of black paper held securely with rubber bands.

If an earthworm is placed on the surface of the earth, it will quickly disappear. A watch with a second hand can be used to time this operation. Later, the black paper can be temporarily removed to get information on the length, direction, and width of the burrow and the position of the earthworm.

Try placing a thin layer of light-colored sand on the surface of the earth in one observation home. Try sawdust in another. Can the children learn about the worm's nighttime activity with the help of these light-colored substances? Is there any difference between the worm's activity in these two substances? What other experiments can the children devise to learn about earthworm behavior and food preferences?

All extra earthworms should be returned to the school ground where they can continue to plow and fertilize the soil and dig holes that help air and water to get down to plant roots.

TEACHER PREPARATION

Examine your school ground for piles of earthworm castings. Castings are hard to find in grass. However, you can be reasonably certain of digging up a worm if you dig into any sodded area.

Trowels or small shovels are always needed for collecting worms, except for specimens stranded after a rain.

FIELD TRIP POSSIBILITIES

1. Take an earthworm census by counting piles of pellets.

2. Collect earthworm pellets and examine them as part of the soil formation story.

3. Hunt for leaves stuck in earthworm holes as part of a soil building study.

4. Find plants growing in earthworm holes where seeds have been pulled down by the worms.

5. Observe dead earthworms on the sidewalk when the sun dries things off after a rain.

6. After or during a rain, watch the effect of earthworm holes on the run-off and run-in of rain.

7. During a rain, take an earthworm census.

8. Do a study of the number of earthworms found in different types of habitats on your school ground.

9. Collect some earthworms for classroom observation or a classroom zoo.

10. Dig up an earthworm burrow by carefully excavating around it. How far down does it go? Does it bend? How large is the chamber (home) at the base?

11. In spring, earthworms of all sizes can be found in moist piles of leaves. Take a census by size and age.

12. Compare the size of mature earthworms (earthworms with a clitellum) found in different habitats. Does quantity or quality of food seem to have any influence?

13. Compare the number of earthworms in a grassy area and a bare area. Discuss the interrelationships involved. How do earthworms help plants? How do plants help earthworms?

A crack in the pavement provides a study area for plant adaptations to limited soil and water supplies as well as the plant's effect on the rock material through the addition of organic substances and the provision of habitats for invertebrates.

Interdependence of Living Things

As we hear reports of scientific discoveries, it seems to many people that we soon will be able to control our natural environment. Not only can we write the chemical formula for DNA we can do genetic surgery so disease should be eliminated. Our space explorations are so mind-boggling that space colonies seem a real possibility if things ever deteriorate on Earth; and as to solid waste, I have repeatedly been told by school children that the solution is easy, send it to the moon.

Many children, and adults, too, live in a science fiction fantasy world. It is time to get our feet firmly planted on Earth. To do this we must build understanding of the beauty and complexity of this unique planet which we inhabit. We cannot afford the arrogance of ignorance.

Even though we have plundered, polluted, and overpopulated the world, we often fail to appreciate the interdependence of all life and of our relative lack of importance. If you doubt this, try asking a group of people to imagine a science fiction situation where all life on Earth is destroyed except for one group of organisms. Ask the people to name the group which could survive longest without the support of any other living things.

In the past year, I have asked this question more than a score of times to grade school children, high school students, college groups, and adults in both rural and urban settings. Except for a college botany class, the answer was unanimous; sometimes hesitantly, sometimes loud and firm, but always "man." In the botany class, where the question was somewhat loaded, the vote was split between "man" and "green plants," but the debate was short-lived when the green plant adherents challenged the "we can always synthesize it" group to take over the green plants' role of keeping the atmosphere of the whole world replenished with oxygen, or to imagine a food store with all the plant and animal products removed from the shelves, or even to synthesize unimportant things in terms of survival like textiles and plastics without cellulose and alcohol from plants.

Finally, when green plants had been agreed upon, it took a while to realize that the survival of green plants varied. A plant dependent on insects for pollination would die without offspring. A plant dependent on animals for seed dispersal would be equally handicapped. In fact, as the semester progressed, it began to be obvious that any kind of flowering plant would suffer without animals to keep the soil turned and aerated, and without other plants to help build soil, hold water, and check erosion.

Finally, the class decided that if only one thing were to survive in our science fiction world, it would have to be those primitive plants that even today are the main source of oxygen for the world, some kind of microscopic algae that grows in the oceans of the world, the kind of plants that were a part of "in the beginning" in the story of life on the planet Earth.

For a while, there were students ready to claim that these plants could go on forever. After all, they could take carbon dioxide and water in the presence of sunlight, produce sugars and starches, and then release oxygen. But protoplasm is not built of sugar and starch. Living things are made of protein. They contain nitrogen. After the algae of the science fiction world had used the available nitrogen compounds, then what? Without nitrogen-fixing bacteria to make new nitrogen compounds, without other bacteria and fungus plants to break down the bodies of algae that got old and died, without animals to feed on the plants and in the process release carbon dioxide and water, sooner or later the plant building blocks, the carbon dioxide, the nitrogen compounds, and the minerals, would be bound in dead bodies. With recycling stopped, nothing would be available for the new organisms.

Even green plants, the producers of the world, cannot stand alone. In order to survive, they need the consumers (the plants and animals that obtain their food from plants or from other animals that feed on plants), as well as the decomposers (the plants and animals that live on dead tissues and break them down).

The interrelatedness and interdependence of living things is undoubtedly the basic concept upon which all our activities and programs must be conducted if living things are to survive on this planet.

Almost every week we read reports of new onslaughts on the environment. Lake Erie, until the 1950s, was an important source of commercial fish. Then phosphates and other pollutants dumped in the lake by industry and sewage disposal plants upset the natural balance of materials in the water. For a while, green plants flourished on the high-phosphate environment. They crowded other things out. They changed the environment so other things died. The bacteria that cause decay increased in number. They broke down dead bodies. In doing this, they used oxygen. Oxygen was already depleted by the chemicals in the lake. It was being used by the plants, too, in the oxidation of food. As the oxygen in the lake decreased, more and more living things died. *Lake Erie was becoming a water desert*. The question became "How can we save this precious resource?"

Scientists, politicians, industry leaders, and private citizens in the United States and Canada worked together to bring about change. Though there is still clean-up to do, people can now swim in the lake, and fishermen are now catching fish.

Lake Erie is an ecosystem (a unit of the environment where non-living substances, producers, consumers, and decomposers are interrelated in such a way as to produce a functioning whole). Coral reefs are also ecosystems. The stony nonliving structures built by coral animals provide shelter and breeding places for a great variety of animals. The shallow waters provide an excellent growing area for microscopic plants.

In the 1950s, starfish began destroying Pacific coral. The ecosystem was destroyed. Fish, which lived on small animals, which in turn lived on smaller animals and plants, which finally, were dependent on the environment created by the corals, decreased in number and disappeared.

What had happened to make the starfish, which had always been a part of the ecosystem, suddenly reproduce at a rate that enabled them to destroy the total environment? Scientists studying the area discovered that snails that fed on starfish eggs had almost disappeared. We are not certain of all the causes, but we do know that just prior to this, the collecting of these snails and the selling of their beautiful shells had become a very important business. Certainly, when the business began no one envisioned the ultimate destruction of the fish which are an essential source of protein to many people living on the islands and lands near the Pacific Ocean. If anyone questioned the uncontrolled snail harvest, the answer would have been, "Why worry about those snails? They're not important. They only eat starfish eggs!"

Oil spills have become common events, and concern for birds and marine mammals trapped by the oil has become widespread. But to the ecologist, an even greater worry is the fate of millions of microorganisms (marine plankton). If these tiny plants and animals are killed, then another ecosystem has been destroyed. Furthermore, oxygen supplies for the whole world will be depleted. The giant ecosystem, Earth, cannot survive the destruction of its interdependent parts.

RELATED CLASSROOM ACTIVITIES

News items can be a good starting point for a study of interrelationships in upper elementary classes. Do research on environmental issues such as the Everglades jet port, the sea-level canal through Central America, and the use of persistent pesticides, to try to understand how a web of life is at stake.

Look around your own community and in your local newspaper. Can you find healthy ecosystems, ecosystems that have been destroyed, ecosystems that could be rescued from destruction?

A study of this type could become an interdisciplinary project involving practically every subject in the elementary curriculum.

In both lower and upper elementary grades, small ecosystems can be set up in the classroom using both terrariums and aquariums. The smallest ecosystem can be created in a corked test tube with an aquatic snail and green plants.

Regardless of whether the consumer (the snail) and producer (the plant) are in balance in their nonliving environment (the water, dissolved minerals, and gases), or out of balance, much can be learned by observing the things that happen.

TEACHER PREPARATION

Teaching interrelationships means looking beyond a single entity to all the things it influences and that influence it. Some plants and animals have obvious two-way relationships, others are highly involved, or very subtle.

For example, a tree may provide food and a nesting site for birds and/or a squirrel. Either of these animals may distribute the tree seeds, providing for continuation of the tree species. Woodpeckers may control the insects that are feeding under the bark; other birds help control insects that feed on the leaves. Insects may help pollinate flowers. They also provide food for the birds. Insects convert dead wood to soil materials, thus making new soil available to new trees and plants.

Dead leaves beneath the tree provide food for earthworms which enrich the soil and make it better for tree growth. An abundant earthworm population provides food for some birds. A fast-growing tree in earthworm-enriched soil produces more leaves, more fruit, more seeds, more bird food, more bird shelter, more earthworm food, more earthworms, and so the story goes *ad infinitum* once you begin to realize all the things that are related to any individual plant or animal.

FIELD TRIP POSSIBILITIES

1. Look carefully around a tree and list all the animals and plants that are in, on, beneath, and around it. Back in the classroom, challenge children to see how many interrelationships they can work out. Can they make charts showing the weblike quality of life in this small ecosystem?

2. Examine a hedge in terms of the influence of producers, consumers, decomposers, and nonliving factors and their many influences on each other.

3. In spring or fall, examine a pile of leaves. What effect have they on the nonliving environment? Look for the white threadlike patterns of fungus plants (decomposers). Can you find animals living on the leaves? Animals living on animals that live on the leaves? How do these animals affect the soil? Are green plants growing in, under, or on the pile of leaves? What conclusions can children draw about this? Do the leaves have any influence on plants around the edges of the pile?

4. Have the class examine a tree stump for interrelationships.

5. A piece of board that has been on the ground a long time frequently has a high population of animals and plants living in and under it. After the children have taken a census, be sure they return the board to its place so the life patterns can go on.

6. In the fall, turn over the soil in two 1-meter-square plots of land that have the same exposure to sun and rain and have the same soil composition. Work dead leaves into one and scatter a single layer of leaves over the surface. Do not add any leaves to the other. In late spring, compare them in terms of plant and animal activity.

7. Let children do a survey of the school ground to see how many sites they can discover where they can explore the interrelationships of living things. Can they then relate the sites to each other? For example, do the birds that are nesting on the school ground because of a tree have an influence on the lawn? Did the presence of the lawn influence their choice of the tree?

8. The interrelationships of insects and flowers are easily observed. If small groups of children stand close to a bush or other flowering plant, they will be able to see a variety of insects coming for food. As long as the children remain inactive, the insects will continue to feed and will not harm the children in any way. Some of the insects will be feeding on pollen, others on nectar. While bees are the best pollinators, many insects carry on this function, thereby enabling the plant to make seeds. This insect activity is not only essential to the plant, but plant reproduction is essential to the insect. For instance, nectar and pollen are the only food of both larval and adult bees. Nectar is one source of adult digger wasp food. Digger wasp larvae eat grasshoppers, which feed on green plants, which grow from seeds their parents helped the plant produce. Caterpillars, the young of butterflies and moths, are also dependent on green plants. Wasps, bees, butterflies, and moths are common insects that may be observed feeding in flowers. Children may discover others that are involved in the story of pollination and benefit from it, or they may find insects and other invertebrates that feed on the feeders. In classroom discussion, they can not only develop the idea of the weblike interdependencies but add themselves to it— for example, without bees and other pollinators, our food plants would be limited to grasses.

9. When two organisms have a relationship that is obviously beneficial to both, it is called mutualism. The bacteria and protozoa that live in a cow's stomach and break down cellulose could not live anywhere else; conversely, the cow would die if it did not have these microorganisms to process its food.

If the school ground has members of the bean family growing on it, children can use them to study another example of mutualism. The bean family is a large family including alfalfa, clovers, vetch, beans, peas, and peanuts. Their seeds are some of the best plant sources of protein. Protein formation requires nitrogen compounds. Special bacteria, which are capable of taking nitrogen from the air and forming nitrogen compounds which plants can use, live on the roots of these plants. They obtain their food from the plant while they provide the plant with the compounds which enable it to produce protein. If the children carefully dig up a clover plant in a lawn, or one of the many wild members of this family in a weed patch, they will discover swellings on the roots in which the bacteria live. Sometimes members of this family are planted to enrich the soil. Children could do some experiments planting two sections of lawn, one with clover and one without, to see the effect of this plant partnership.

They may think of other experiments they can do if they ask the question, "What would happen if . . . ?"

10. The group of plants known as lichens are another example of mutualism. Here a fungus and an alga are so interwoven that they appear as one plant. The fungus cannot make food and obtains it from its green plant partner. This combination plant is an important soil builder. As the underpart of the lichens age, they die, decay and add organic material to the mineral soil on which they grow. The fungus also secretes substances which dissolve rock, thus making minerals that other plants can use available in the newly formed soil. Look for lichens on stone walls or branched lichens growing on waste soil. Many lichens are pale green, a mixture of white fungus and the chlorophyll of the alga. Sometimes they contain other pigments, such as black, brown, or orange, which cover the green and white. (Note: lichens will not be found in areas of high air pollution.) Observe them before and after a rain. Notice how the fungus part serves as a sponge.

11. A story of mutualism which can be exciting for children to observe is the story of ants and aphids (plant lice). Aphids feed on plant sap. When they

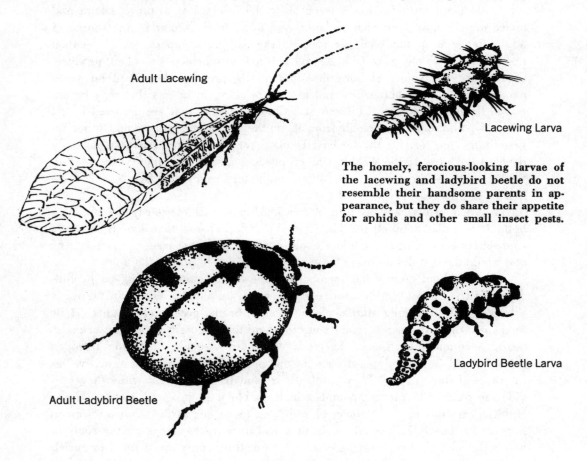

Adult Lacewing

Lacewing Larva

The homely, ferocious-looking larvae of the lacewing and ladybird beetle do not resemble their handsome parents in appearance, but they do share their appetite for aphids and other small insect pests.

Adult Ladybird Beetle

Ladybird Beetle Larva

build up a sap surplus in their bodies, they secrete a special sweet substance known as honeydew. Ants use honeydew for food. When an ant taps an aphid with its antennae, the aphid secretes the honeydew. The ant drinks the honeydew and moves on to the next aphid. Can you see why aphids are called ants' cows? When the ant's stomach is full, it hurries back to the ant colony to feed ant larvae and stay-at-home ants by regurgitating the honeydew. If an aphid enemy, like a lacewing or a ladybird beetle, comes to the aphid colony or lays her eggs in the colony so her hungry babies hatch and feed there, ants often gently pick the aphids up in their mouths and carry them away from danger. Sometimes all of these activities can be observed in one aphid colony at the same time. Usually at least one activity can be seen. In fact, ants on plants often are an indication that aphids are not far away. A few kinds of ants have even more complicated relationships with special aphids. After children have observed the common types of relationships for their area, some of them may enjoy doing some library research on things like corn aphids and ants, "ants' cow-barns," or ants that become "storage jars" (repletes).

Characteristics of Living Things

A maple tree, a bread mold, an eagle, an amoeba, a termite, an earthworm, and a person are only a few examples of living things. The list could go on and on, yet it wouldn't simplify the job of defining living. How do we separate the living from the nonliving, the alive from the dead?

We look at a dog struck by a car on the street and we say, "He doesn't move, he's stiff, he's dead, he's not alive."

We look at a lichen, which looks like a coat of gray paint on a rock. It doesn't move. It is stiff. But we say it's alive.

We walk up to the door of a supermarket which is controlled by an electric eye. The door moves. It swings open as we approach, but it is not alive. Obviously, this special quality or state that we call life is not easily described.

All living things are composed of a substance called protoplasm. Protoplasm has the ability to do certain things. It takes in food and combines the food with oxygen to obtain energy. This energy is used to carry out a variety of activities. We describe these activities as characteristics of life.

One of the characteristics of life is growth. All living things start life small and finally arrive at the size and shape that is characteristic for their species (kind of plant or animal).

All living things have a definite size or shape. Mice do not grow to elephant size. A Norway maple tree has a different size and shape from a eucalyptus tree, a geranium, or even a sugar maple tree.

Living things have the ability to repair their own tissue. This varies from the ability to heal cuts and broken bones in vertebrate animals; to the ability to grow new parts like tails in lizards and earthworms, legs in arthropods, and branches in trees; to the ability to grow whole new organisms from small pieces as is done by flatworms, starfish, and many plants.

Living things reproduce and pass on their particular qualities to their off-spring. Techniques of reproduction vary from the production of a single cell called a spore which will grow into a new individual in many kinds of plants, and some animals; to breaking in half to form two individuals in some plants and animals; to egg-laying; seed production; and giving birth to living young. Regardless of technique, the end result is the carrying on of the species.

This is important because all things composed of protoplasm age and finally die. The potential age for any plant or animal is an inherited character-istic of that species. Many insects have a one-year life span. They will not live longer than one year. They may live less because of starvation, parasites, predators, storms, or other accidents. Sequoia trees have a potential life span that runs into the thousands of years but they, too, age and die.

Mice, gerbils, and hamsters all have a two- to three-year life span. Their death from old age is a normal part of living.

In recent times, we have seen an increase in people's life spans in some parts of the world simply because with improved diet, parasite control, and the reduction of other causes of bodily injury, more people are approaching the potential human life span.

Living things respond to stimuli. Responses range from growing toward light in plants and moving away from light in nocturnal animals like earthworms, to a complicated variety of reactions to a large number of stimuli in vertebrate animals and, particularly, in humans.

Living things adapt to their environment. This may include plants growing longer roots in desert areas where water is not easily available. In humans, adaptation can include physical changes such as the development of an enlarged chest cavity, heart, and lungs by people living in very high altitudes where oxygen supplies are low, or mental responses such as tuning out disturbing sounds.

Living things have specialized parts which enable them to carry on these and other specific functions. We call these parts organs, and we call living things organisms.

The dead dog was composed of protoplasm but it had lost the ability to carry on the functions that characterize life. Ultimately, bacteria and other or-ganisms break down the protoplasm to the simple nonliving substances from which it was built.

The lichen is composed of protoplasm. Like many plants, its motion is re-stricted to growth and to expanding and contracting in response to water. But observation will reveal that it does carry on all the life functions.

The supermarket door is made of metal—it isn't alive and never was. Its motion comes when an electric circuit is closed either by a photoelectric cell that responds to your shadow or by the pressure of your foot on the door mat.

Motion and other responses in living things are initiated from within the organism. Motions of doors, automobiles, kites, and many other things are not self-initiated.

RELATED CLASSROOM ACTIVITIES

The study of the characteristics of living things may grow out of almost any classroom experience with plants or animals.

Brine shrimp and planaria are two kinds of invertebrate animals which demonstrate growth, reproduction, response to stimuli, and repair extremely well in a short period of time. They may be obtained from pet shops and biological supply houses.

Growth and reproduction in plants can be dramatically observed through experiments with bread mold.

Experiments with asexual reproduction (reproduction involving only one parent) may be carried out by trying to root coleus and other plants in water, vermiculite, or sand.

TEACHER PREPARATION

Any school ground provides an opportunity for at least one field trip about living things, even if it is a negative trip to discover the factors that prevent life from existing there. Most school areas will provide more than this. In fact, you may be surprised by the variety available once you become sensitive to your school site.

FIELD TRIP POSSIBILITIES

1. On a nice day, go out and enjoy the sensation of being alive. Have the children close their eyes and use their skin, ears, and noses to make discoveries.

2. See how much your class can learn about an area using only one sense.

3. Select a plant on the school ground, and watch it over a period of time to see how many characteristics of living things you can observe.

4. Have the class observe a plant like a maple tree and an animal like a sparrow. See if they can find ways in which both illustrate the characteristics of living things.

5. Observe a variety of plants and animals. Make a chart showing how each grows, repairs, reproduces, responds to stimuli, and adapts. Can the class make some generalizations about the similar behavior of plants and animals? For instance, all living things respond to stimuli. Motion is one type of response in living things. Compare the motion of most plants to the motion of most animals. Can the class draw some conclusions about plants and animals?

6. Examine as many of one species of tree or other plant as you can find on your school ground. Observe the fairly definite size or shape which characterizes the species. Notice how much alike the size and general shape of adult trees of one species growing in full sun are. Find ways in which young trees are like their parents: similar leaf arrangement, leaf shape, leaf color, and branching.

7. Observe a flock of one species of birds. How do you know they are one species? Are they alike in ways other than size and shape? Can you discover differences between males and females?

Male House Sparrow

Female House Sparrow

House sparrows are excellent birds to study because they will completely ignore quiet observers and go about their business. The black-bibbed males are easy to distinguish from their duller-colored mates.

8. Look at two trees of the same kind, one growing in full sun, one in an area shaded by the building. Can the class discover any adaptations the shaded tree has made in response to its environment?

9. Examine a tree trunk for scars that indicate that the tree has repaired injuries. Can you find places where branches have broken off that are now partly or completely covered with bark?

10. In the spring, go outside and try to find ways in which small living things are reproducing. Look for eggs of insects, spiders, and other invertebrates on tree trunks or bushes, under rocks or boards, under building trim, or in other protected spots. You may find sowbugs and pillbugs carrying their eggs on the underside of their bodies. Look for seeds on plants. Can you find vines or bushes like forsythia that are making new plants by growing roots where the ends of branches touch the ground?

Requirements for Life

Living things can survive in an amazing variety of environments and can often withstand great discomforts as long as they can obtain oxygen, water, and food. Many organisms have developed special structures or techniques (called adaptations) for procuring these substances which are essential to life.

Regardless of whether a molecule of sugar, protein, or fat is within the plant leaf where it was produced, or in the stomach of an animal that has eaten the plant, or in the tissues of a fungus plant that has absorbed it from the dead body of a plant or animal, it is useless until it has been combined with oxygen.

Air enters the leaves of flowering plants through microscopic holes on the leaf surface called stomata (singular: stoma). Inside the leaf, carbon dioxide is removed from the air and combined with water to make food in the process called photosynthesis. The chloroplasts perform photosynthesis only in light and within a certain range of temperature varying according to the climate.

Oxygen is also removed from the air which enters the leaf through the stomata. Within the leaf cells, it combines with the food to release the energy that was stored by photosynthesis.

This is essentially the same process which occurs in the cells of animals. Vertebrate animals have nostrils or gill slits, large openings through which air enters their bodies. Insects have paired holes called spiracles on each abdominal segment. Snails and slugs have a single spiracle. Regardless of how air gets inside the organism, oxygen is removed and travels to every cell. Sometimes it is carried by the blood. Sometimes it is passed from cell to cell. In many small aquatic animals and some land animals with moist skins, like earthworms and red-backed salamanders, the oxygen enters the cells and/or blood directly through the skin.

Methods of obtaining water are just as diverse as methods of getting oxygen into the body. Some animals drink water. Some drink dew from plants. Some obtain their water from succulent leaves of plants, from juicy fruits, or from the body fluids of the animals on which they feed. Most flowering plants obtain their water through roots. In dry areas, a small plant may have roots that go down 6, 9, or even 12 meters.

In addition to being used in food production by plants, water is used by all living things for equalizing temperatures, bathing protoplasm, transporting

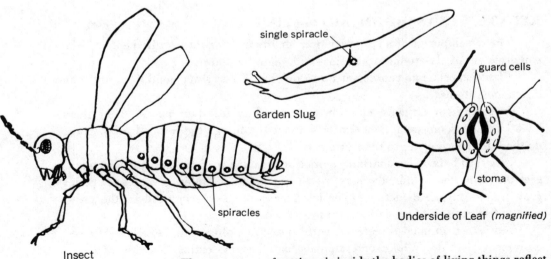

single spiracle

Garden Slug

guard cells

stoma

spiracles

Underside of Leaf *(magnified)*

Insect

The many ways of getting air inside the bodies of living things reflect the great variety of life on earth. In insects, spiracles lead to a complex internal network of tubes; in slugs and land snails, the spiracle leads to a single lung; in plants, guard cells control the size of the stoma (opening) which permits air to enter and leave the leaves.

food, and removing waste products. Blood and sap are both largely composed of water, as is all protoplasm.

While all living things have built-in techniques for getting wastes out of their bodies, disposal of these wastes can be critical. This is particularly true of animals that live in a restricted environment. Some leaf miners always return to one end of their mine to defecate. Miners that are caterpillars use silk to fasten their droppings down and guarantee that they remain behind them. Frequently, a gall insect feeds on the upper walls of a leaf while the bottom of the inner chamber is filled with its droppings.

The same kind of pattern of feeding and defecating can be observed among pasture animals. Droppings may be concentrated in one area where the animals congregate to rest and chew cuds. Where droppings are scattered over the field, tall clumps of grasses often mark places where the animals grazed around a pile of manure. This instinctive avoidance of their own wastes helps reduce infection from intestinal parasites.

Birds clean house by throwing the droppings overboard every time they bring food to their young; and as soon as the young are strong enough, they are taught to sit on the edge of the nest and defecate outside it (a habit that may make a bird lover who has permitted a bird to nest on his porch or windowsill very unhappy, but one that materially reduces infant mortality in birds).

In addition to a clean habitat, living things need a safe habitat. This means it must provide protection from enemies and shelter from extremes of weather. (Plants and animals all have their own optimum temperature and climatic conditions.) If the species is to continue, it must be a habitat where the organisms are able to rear offspring.

RELATED CLASSROOM ACTIVITIES

The development of an appreciation for life and for the requirements of life can grow out of classroom experiences with plants and aimals.

Plant experiments can be set up to determine necessary conditions and optimum conditions.

In caring for animals, children are providing for their needs. After they have had classroom pets, they can talk about the things the pets all need. How do these things compare to their own needs?

In a well-balanced aquarium, animal wastes are recycled and used in plant growth. Too many animals may result in polluted water, overactive plant growth, and oxygen depletion. (The latter can often be recognized by the presence of fish at the surface gasping for breath.)

Stories of animals threatened with extinction often can be related to requirements for life. While some animals have become extinct because of overhunting, many others have been eliminated by the depletion of food supplies or the destruction of their habitats. In today's world, the depletion of food supplies sometimes includes the poisoning of foods with substances like DDT and mercury.

TEACHER PREPARATION

Almost any living thing that can be found on the school ground can be used as a basis for exploring the requirements for life. By the same token, a dead organism may raise questions in children: "Why did it die? What requirements for living were lacking?"

FIELD TRIP POSSIBILITIES

1. If a bird is nesting on your school ground, do a study to see where it gets its food and water. What shelter does it use for rearing its young? How does it obtain protection from weather? Why is it important not to get too close to the nesting site?

2. Green plants use the energy from the Sun to produce food. Compare two trees of the same age, one growing in shade and the other growing in the Sun. Is there a size difference? Can it be related to food?

3. Plant two groups of plants of the same kind and size, one in Sun and one in shade. Observe their growth.

4. If your building has an overhang, examine the area beneath it for plants. What requirement for life is missing?

5. Have the children take a survey of the areas on the school ground where plants do not grow. Decide which requirements for life are missing.

6. Sometimes trees next to a driveway or along a curb die because air containing carbon monoxide enters the leaf stomata and displaces the oxygen. Sometimes a comparison of the size and general health of trees in these places shows that even trees that don't die are not healthy. A map showing the location of healthy and unhealthy trees along a curb or driveway may be correlated with the places where buses habitually stop with motors idling.

7. Look at trees in the sidewalk or on a hard-topped play area. Discuss the area around the trunk. Many kinds of trees are unable to survive on the limited water available in this type of unnatural habitat.

8. After you have discussed the requirements for life in the classroom, plan a school ground census. Let children predict the relative number of animals they think they will find in any given area on the basis of these requirements. Go outside and see if the areas measure up to their predictions. If not, can the children discover things they overlooked in working out their predictions?

Microhabitats

A microhabitat is a small unit of the environment especially suited to the growth and development of an organism or a group of organisms. In its narrowest sense it may be a small home, in its broadest a small ecosystem.

It is frequently associated with a microclimate. The term microclimate refers to the climate of a small unit of land which is different from the total climate of an area. Central Park in New York City has a microclimate. Its trees, grass, soil, and bodies of water influence the pollution level, the temperature, the winds, and the effect of precipitation.

The little strip of land along the south side of a building where heat absorbed by the wall is reflected on the ground provides a microclimate which sometimes permits the growing of plants that do not ordinarily survive in the area. For instance, there is an Italian garden in the Bronx where tomatoes, oregano, marjoram, peppers, melons, and fig trees flourish. It is maintained by the residents of the apartment house who recognized the south-facing wall as a microclimate that could be converted into a microhabitat to provide both pleasure and good eating.

The first family in a town or city who have crocuses or snowdrops blooming every year don't have a green thumb; rather they have a sensitivity to the environment which enables them to recognize warm microclimates and convert them into microhabitats for early blooming plants.

Truck farmers treat their farms as a series of microhabitats. One successful New England farmer raises apples on his hills. He says, "Frost settles in the valleys," so apple blossoms on trees in the lowlands often get killed while the ones higher up survive. He has the first sweet corn for sale in the area because one of his fields has sandy soil, which he can plow and plant sooner than fields with clay soil, which remain wet until much later in the spring. The first of these farm microhabitats involves a microclimate; the second is based on soil composition.

Sometimes the inhabitants of a microhabitat make a microclimate. A gall is one example of this. Galls are structures produced by plants in response to enzymes deposited on the plants by animals or bacteria.

A common gall is the goldenrod ball gall. In the early spring when the goldenrod plant is young, a fly with brown designs on its wings lays an egg on

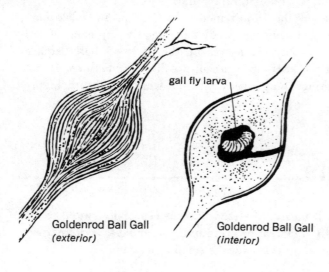

gall fly larva

Goldenrod Ball Gall
(exterior)

Goldenrod Ball Gall
(interior)

Galls can be found wherever plants grow. The goldenrod ball gall is one of the most widespread.

the goldenrod stem. The tiny white grub that hatches from the egg bores into the stem and secretes a substance which stimulates the stem to grow into a sphere about an inch and a half in diameter. At that point, this makes a perfect home for the little grub; its soft body is protected from drying winds, driving rain, and hot sun. The moist plant tissues provide food and shelter from enemies as well as a perfect indoor climate.

There are three other kinds of galls which are found on goldenrod. Indeed, galls grow on practically every kind of plant but are most common on oaks, willows, roses, blackberries, and goldenrod.

Any yard or school ground will have many microhabitats. Whenever a home-owner who enjoys gardening or an apartment dweller interested in house plants asks, "Where will this plant grow well?" he is seeking a microhabitat appropriate to the needs of one living thing.

RELATED CLASSROOM ACTIVITIES

A terrarium or aquarium is a classroom microhabitat. In can be extremely interesting to have several terrariums representing different habitats like woods, bogs, deserts, and grasslands. To be successful each must be planted in soil from the appropriate area so that chemical and physical properties of the environment are correct.

The classroom itself may provide several microhabitats. In fact, the topic could be introduced by trying to find the best spot to put some organism. Is the whole room equally good for green plants? For hamsters? For the growth of mold?

TEACHER PREPARATION

A square building has four different exposures to the sun; with extensions, ells, overhangs, and other supplementary structures, the variety of microclimates and microhabitats increases. In fact, two south walls, one shaded by a perpendicular wall and the other jutting forward, can provide excellent side-by-side areas for experimentation and observation.

FIELD TRIP POSSIBILITIES

1. In the spring, let the children search for insects that make their own microhabitats. In addition to galls, this may include spittle bugs, which extrude a liquid and beat it into froth; leaf rollers, a kind of caterpillar which folds leaves over its back and sews them down with silk; and leaf miners, larvae of tiny beetles, wasps, flies, and moths which live between the top and bottom layers of leaves and feed on the inner cells. Decide how these habitats are different from the surrounding environment. How do they provide food, shelter, and protection from enemies?

2. An anthill in a crack in the pavement or asphalt play area is an indication of another microhabitat. The hill is made of sand and soil particles

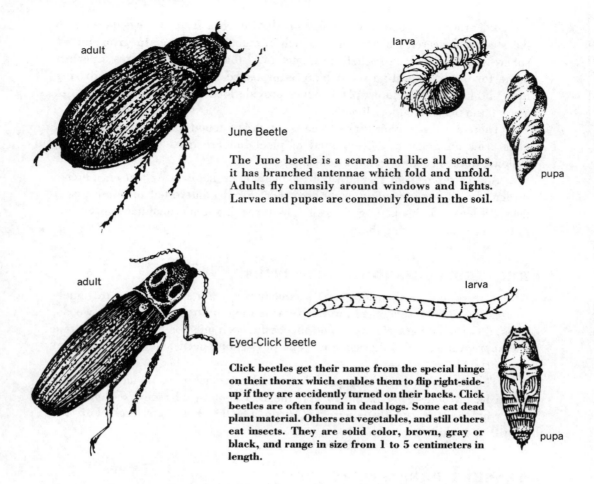

adult

larva

June Beetle

The June beetle is a scarab and like all scarabs, it has branched antennae which fold and unfold. Adults fly clumsily around windows and lights. Larvae and pupae are commonly found in the soil.

pupa

adult

larva

Eyed-Click Beetle

Click beetles get their name from the special hinge on their thorax which enables them to flip right-side-up if they are accidently turned on their backs. Click beetles are often found in dead logs. Some eat dead plant material. Others eat vegetables, and still others eat insects. They are solid color, brown, gray or black, and range in size from 1 to 5 centimeters in length.

pupa

brought to the surface from underground galleries. What advantage might the area under a pavement provide for these animals that live in tunnels? Observe the ants going about their business. What are they doing? Where do they get food?

3. A stump, a dead log, or a piece of board are often small ecosystems. Beetles, beetle larvae, termites, ants, sowbugs, pillbugs, millipedes, and other invertebrates feed on the decaying wood. Other beetles and beetle larvae, centipedes, spiders, and mites feed on invertebrates feeding on the decaying wood. Slugs and snails may hide in the holes in the moist interior. Once the wood has been broken down by insects, bacteria, and fungus plants, earthworms will begin to live in the soft material. Moss and lichens may grow on the outside. The number of organisms that are found in a microhabitat of this type will be related both to its size and to its age. After the class has examined a decaying wood microhabitat, the wood should be returned to its original site if at all possible so it will continue to be available for future study (and soil enrichment).

4. Children may be able to find several microhabitats on a rock outcropping or a stone wall. Are there plants growing on the bare rock surface? Are there

plants growing in the cracks? Are there differences in these plants? How do the habitats differ?

5. The area under a rock is often a good place to discover a variety of animals and animal relationships. Like the dead log, this microhabitat provides shelter and considerable environmental control. It differs from the dead log in not being edible, and in being relatively unchanging for a long period of time. It is easy, and important, to return it to its original position.

6. Examine tree trunks for moss and lichens. Do they grow all around the tree or only on some parts? Students might draw the tree trunks from two sides and show where the moss is growing. Watch the tree trunks at different hours of the day. Does the sun ever shine on them? Students might make a map of the part of the tree that gets direct sunlight. Compare it to the map showing moss growth.

If the trunk is never in direct sunlight, let the children use a light meter to record the amount of light on each side of the trunk at hourly intervals during the day. From these experiments, can they draw conclusions about the effect of sun on this microhabitat?

7. Often galls which started out for one kind of animal become a microhabitat for other animals. This is particularly well illustrated by the pine cone willow gall which grows on all kinds of willow trees. The gall is made by a gall fly. The fly larva is an orange grub which lives in a cell in the center, but other insects lay their eggs beneath the scales. Some of the young insects also feed on the gall. Others, like baby grasshoppers, move out into another habitat. Even after the original owner-builder of the gall has emerged, the old galls often continue to provide a microhabitat for other insects. Collecting and taking pine cone willow galls apart can provide a variety of experiences. Even if the galls turn out to be empty, the way in which the plant grew in response to the insect-deposited enzyme is extremely interesting. (Each scale would have been a leaf on a branch if the gall-maker had not interfered to create a microhabitat.)

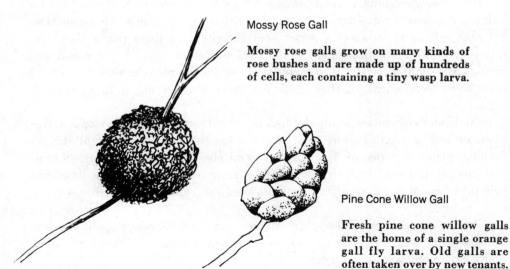

Mossy Rose Gall

Mossy rose galls grow on many kinds of rose bushes and are made up of hundreds of cells, each containing a tiny wasp larva.

Pine Cone Willow Gall

Fresh pine cone willow galls are the home of a single orange gall fly larva. Old galls are often taken over by new tenants.

8. Often a hole in a tree trunk serves as a microhabitat for a variety of plants and animals. The inside part may show evidence of its return to soil materials both in texture and by the plants growing in it. It may be a habitat for a bird, a mouse, a squirrel, or other vertebrate animals, or it may contain invertebrates. Examining the habitat in terms of size, moisture, location, protection from weather, and consistency of the wood may help the class to predict the living things that might be found there. Discovery of signs such as droppings, nesting materials, shells cracked by squirrels, shells with round holes gnawed by mice, and stored food supplies will confirm the presence of some living things.

9. Often, the study of microhabitats can help build an understanding of macrohabitats (large habitats). A hedge is a microhabitat with many of the properties of a forest: dense shade, soil holding, soil production resulting from the dropping and decay of leaves, effect on rainfall, and shelter for birds. It can be a good observation and study site for a class that is discussing forests.

10. Lawns, too, are microhabitats similar to the grassland macrohabitats. Much of the power and importance of grasslands in holding soil and water and serving as an active ecosystem with all kinds of producer, consumer, decomposer, and nonliving substance relationships can be learned by observation and experimentation with a lawn.

11. While an anthill in a sidewalk crack is the story of the microhabitat under the paving blocks, the cracks between paving blocks are microhabitats in themselves. Mosses, grass, weeds, even plants like Ailanthus trees can be found growing there. A class could do a survey of pavement-crack microhabitats. Are there differences between the places where moss is growing and young trees are growing? Can the children find places where the plants have affected or enlarged their habitat? Can they find places where plants have outgrown their habitats? Is there ever a relationship between anthills and plants? Does the survey turn up any other pavement-crack animal stories?

12. Good gardening is really taking advantage of and/or making a microhabitat. Suppose a class decides to raise radishes or marigolds. To be successful, they will need to select a sunny spot for either of these plants. Each requires certain soil conditions. If these conditions do not exist, they can add leaf mold, peat, sand, lime, or other materials to make the microhabitat the correct one. What techniques can they work out for providing the right amount of water?

13. Upper elementary grade children in a rural setting could develop a native plant or fern garden. This requires much knowledge of the needs of the individual plants in terms of light, physical and chemical properties of soil, and moisture. It is foolish and destructive to plant a fern that requires limestone soil in a bed of peat, or a fern that requires the acidity of oak leaves on a ledge of lime.

Young people can derive great satisfaction from developing appropriate microhabitats and enjoying the success of their venture in the form of a flourishing microbotanical garden.

Population Explosion

The reproductive potential of any living thing is a reflection of that organism's niche in the environment. Codfish congregate at the ocean surface and discharge eggs and sperm. One female will lay millions of eggs annually during a life span of many years. This means that two codfish could have hundreds of millions of offspring replacing them in the ocean.

Trout, on the other hand, dig nests in the rocky stream bottom. A male and female discharge sperm and eggs into the nest side by side, then the female covers the eggs with rocks and pebbles. Trout eggs number from hundreds each year in youth and old age to thousands at the prime of life for a period of about eight years. The potential replacement for two trout is between 20,000 and 30,000.

This is a huge number compared to field mice, which can breed every three weeks and potentially could produce 17 litters of five to nine babies for a total of 85 to 150 offspring for two mice in a year. (The females generally die of exhaustion in that length of time.)

These and all other animals have a tremendous potential for overrunning their environment. Yet it rarely happens. With the casual extrusion of cod eggs and sperm, some eggs never get fertilized, and the mortality is extremely high for the millions and millions that do. Gulls, terns, other sea birds, and all types of fish congregate to scoop the floating eggs from the surface of the ocean. Many of the tiny fish that hatch from the surviving eggs also fall prey to all kinds of marine animals.

Trout eggs have a much better chance of being fertilized and are well protected from predators. Only parasites and silt get down into the nests, feeding on or smothering the eggs. But the early stages of food gathering and hiding from enemies are extremely perilous for baby trout, and only a few reach their first birthday.

Mice have a much better chance of reaching adulthood, with internal fertilization, maternal care, and a ready food supply. But fortunately for mice and humans, predators (owls, hawks, crows, frogs, snakes, turtles, shrews, skunks, and weasels) generally keep them in check. If they don't, and the mouse population approaches its potential, as it did in Humboldt Valley, Nevada, in 1908, the thousands of mice to an acre eat everything in sight and finally most starve to death.

In fact, disease and starvation ultimately kill off populations that go unchecked. For example, the deer herds in Isle Royale National Park in Lake Superior and in the Kaibab National Forest in Arizona were "protected" from disaster by the elimination of wolves, coyotes, and other predators. In the Kaibab National Forest, 4,000 mule deer increased to 100,000 in 20 years. In the next 15 years, disease and starvation brought the population down to 10,000. The forest became devoid of young plants and trees and gullied by erosion. The ecosystem had been destroyed.

Regardless of whether it is producers, like the algae in Lake Erie, or consumers, like the starfish on the coral reefs or the deer of the Kaibab, or even decomposers, like the mildews and molds in a too-wet environment, any population explosion spells disaster for both the species and the habitat.

How does this apply to humans? We, too, are biological organisms subject to natural law. Our reproductive potential, one child per female per nine months for 25 or 30 years, was just as much in ratio to the hazards of the ecosystem a hundred thousand years ago as was that of the cod, the trout, the mouse, and all other living things. Conditions have changed, however. Saber-toothed tigers, wolves, cave bears and other predators were eliminated as population controls long ago.

In the past century, many human parasites have also been eliminated. In the early 1900s, a popular ballad told of "the little white hearse" carrying children to the cemetery, children killed by diphtheria, polio, tuberculosis, scarlet fever, rheumatic fever, and other childhood diseases. Today almost no family must have five or six children in order to rear two.

But today we are threatened by both the psychological and physical controls that accompany overpopulation. As increasing numbers of the world's people need food, water, and shelter, we need to apply our knowledge of checks and balances and the interrelatedness of life to ourselves. We must somehow stop our population spiral before we, like the deer of the Kaibab and Isle Royale, totally destroy our environment—the ecosystem, Earth.

This does not mean restoring human predators and parasites, or moving out into the hostile environment of sun, sleet, snow, and storms unclothed and unhoused. It does mean, however, that without looking backward either with nostalgia or blame, we must accept the changes that have occurred and as individuals take responsibility for population control, so that our grandchildren and great-grandchildren will inherit an inhabitable ecosystem.

RELATED CLASSROOM ACTIVITIES

Do some arithmetic studies of potential population growth. Start with a pregnant nine-week-old field mouse giving birth to her first litter of six. She will mate a few hours later and three weeks later, she will give birth to another litter. At this rate, when she's 18 weeks old, she will have had 24 children and 18 grandchildren. (Assuming that each litter is half male and half female, three nine-week-old pairs will be reproducing at this time.) She will be a great-grandmother at 27 weeks when her first grandchildren (nine males and nine females) will have 54 offspring. Of course, this assumes no predators, no parasites, adequate food, and a mild climate that permits all-year breeding. If the litters averaged six, the mouse would have 6,206 offspring at the age of 54 weeks. If the litters averaged eight, she would have 21,776 descendants at 54 weeks.

When this kind of population explosion occurred in Nevada because of the absence of natural predators, gulls moved in to harvest the mice that were rapidly eating themselves into a starvation situation and the farmers into bankruptcy.

POPULATION POTENTIAL OF FIELD MICE

Age of Female Mouse (in weeks)	Children	Grandchildren	Great Grand-children	Great, Great Grand-children	Great Great, Great Grandchildren	Great, Great Great, Great Grandchildren
9	6					
12	12					
15	18					
18	24	$3 \times 6 = 18$				
21	30	36				
24	36	54				
27	42	72	54			
30	48	90	108			
33	54	108	162			
36	60	126	216	162		
39	66	144	270	324		
42	72	162	324	486		
45	78	180	378	648	486	
48	84	198	432	810	972	
51	90	216	486	972	1458	
54	96	234	540	1134	1944	1458

Doing the same kind of computation with robins, which are well-liked and reproduce much more slowly, can be equally interesting. Start with a pair of robins in your backyard. As a rule, they raise two broods of four per year. The next year, without natural checks and balances, 10 robins (five pairs) would raise 40 young. The following year, 25 pairs would raise 200 young. At four years, the potential offspring for 125 pairs of robins is 1,000!

Population explosion can be demonstrated using a fruit fly colony which can grow to huge proportions in a very short time. Fruit flies may be purchased from a biological supply house or wild ones may be enticed to feed and lay eggs on overripe fruit like grapes, bananas, or pears in an open container. When several flies have congregated, the container can be covered. Fruit flies have complete metamorphosis. The eggs hatch into tiny white maggots, which feed on the fruit and pupate in three or four days. In another four or five days, adult flies emerge ready to mate and lay more eggs, which will hatch, feed, pupate, emerge as adults, mate, lay eggs, and so on *ad infinitum* until sanitation problems or depleted food supply bring the population under control.

TEACHER PREPARATION

The same areas that provide good sites for the study of seed dispersal will also be good areas to observe population potential and controls.

Any school ground with trees will offer the opportunity to gain an impression of the tremendous oversupply of plant seeds and the marvelous system of checks and balances on the plant population.

Animals will usually be harder to study in this context on the school ground, although flower beds may have aphids (plant lice), which are ideal subjects for a population study.

FIELD TRIP POSSIBILITIES

1. Have children count the number of seeds in a fruit head of a sycamore or London plane tree (children call them "itchy-balls"), then estimate the number of heads on the tree. How many seeds does this represent? Sycamore trees may live to be several hundred years old and reproduce at this rate steadily. Look for young sycamore trees. Sycamore seeds will only germinate in wet soil with a lot of decayed plant material. Even though your tree produces millions and millions of seeds in its lifetime, what are its chances of having one young tree to replace it?

2. If you have a silver maple tree on your grounds, watch its seeds fall in late May or June. Estimate the number on the ground. These seeds die as soon as they dry out, which means they have a maximum life of about two weeks off the tree, unless they reach a suitable spot for germination. In addition, many of these juicy early spring seeds are eaten by birds and squirrels. Look for the maple keys (children call them "Polly noses") with the seed eaten. Birds, squirrels, and environmental factors all keep this tree population under control.

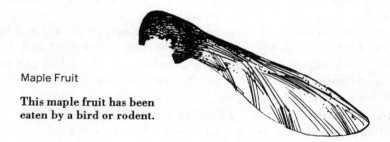

Maple Fruit

This maple fruit has been eaten by a bird or rodent.

3. The seeds of Norway maple trees do not ripen until fall. They lie on the ground during winter and germinate in the spring. Young Norway maple trees may be found under the parent tree, in lawns, flower gardens, and waste places. Make a school ground map in early spring and record the germinating Norway maple trees. Check them every week. Can you find natural checks beginning to work, like overcrowding, shade, animals, or the activities of people?

4. In the fall, take a census of one kind of weed or grass on the school ground. Count the seeds on an average plant. What is the population potential? Watch for things that serve as population controls during the winter. In the spring, look for young plants. Is there an increase or decrease in the plant population? Are any controls acting on the young plants?

5. The most successful city tree is the Ailanthus. Can you find an Ailanthus population explosion on your school ground? If the trees are growing close together, do they show the effects of competition for sunlight, soil, and water? Can you find Ailanthus trees that have died because they have exhausted the nonliving part of their habitat?

6. Examine an aphid colony. Baby aphids look like small versions of their mother. Some adult aphids develop wings but many do not. Notice the different

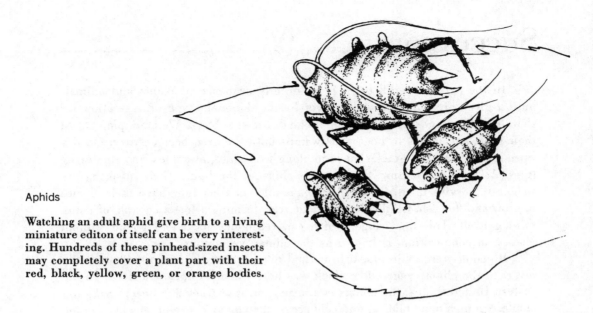

Aphids

Watching an adult aphid give birth to a living miniature editon of itself can be very interesting. Hundreds of these pinhead-sized insects may completely cover a plant part with their red, black, yellow, green, or orange bodies.

sizes which represent different ages. Count the aphids. Look at the plant on which they are feeding. Is it wilted? Is it stunted? If it is, the winged aphids will be able to fly to another plant but the others will die when the plant dies unless ants move them. Look for population controls like ladybird beetles or their larvae, lacewing flies or their larvae, or for empty aphid skins with a hole in the abdomen where a parasite wasp has fed on the inside of the aphid, then pupated inside the empty skin, and escaped.

7. Examine an asphalt or cement section of the playground. How many seeds can you find that have fallen on unsuitable ground? Are they all the same kind? In what other places might seeds fall in the city and never have a chance to grow?

8. Nuts and acorns are distributed by squirrels. An acorn will germinate on the surface of the ground but other nuts must be planted. Watch squirrels storing nuts in trees, in their nests, and in the ground. During the winter, watch the squirrels digging up their caches. Can the class see why the tree must produce large numbers of acorns? Another check on oak population can often be observed either by cracking acorns or by putting them in a glass jar and discovering the white caterpillars or beetle grubs that emerge looking for a place to pupate.

9. Take a field trip to look for evidence of too many people living in a small area. Is rubbish removal partly a problem of numbers? Is grass destroyed by the pounding of thousands of feet? Is there a fence around grassed areas? Can children see why protection of plants from sheer numbers of people is necessary? Might it be better if the people now concentrated in a small area in apartment buildings were spread out over a larger area? Is the school play area and grassed area reduced in size because of supplementary buildings put up to take care of an expanding school population? What would happen to the school if the population is doubled in the area?

Succession

In the natural world, there is an orderly sequence of plants and animals until a given area finally is taken over by the dominant type of vegetation for that locality. Ponds gradually fill up and become bogs. Wetland plants add their roots and leaves to bogs and swamps until the area becomes dry. On dry open land, weeds and grasses move in along with vines, brambles, and sun-loving trees. As these trees grow, their crowns shut out the light. Their offspring are unable to survive in this shade, but shade-tolerant trees now have their chance to grow and flourish. Gradually they take over, forming a forest capable of maintaining itself. This final stage, which can reproduce itself and go on without change in composition, is known as the climax vegetation.

If people were to disappear from the United States, the land would gradually revert to the climax vegetation which was here when Europeans arrived. In the eastern United States, the climax is a mixed forest of trees like beech, oak, and maple. On high mountains and in cold areas, the climax is evergreens like spruce and fir. In the central United States, grasslands represent the climax vegetation. Mixed forest, evergreen forest, chaparral (low thorny shrubs), and temperate-zone rain forests each represent climax forms in one western area or another. Climax vegetation for any area is determined by available light, moisture, temperature, and soil types.

Although the series of changes involved from open lake to climax forest takes hundreds of thousands of years, the process of natural succession can be observed in small units of time and space almost anywhere.

A stone wall in the woods of any New England state tells many stories. Stone walls were built of rocks removed from fields when land was under cultivation. As better farm land became available in the west in the nineteenth century and the Industrial Revolution enabled New Englanders to earn a living by harnessing their rivers for factories, farms were abandoned.

Unplanted fields were taken over by weeds, grasses, brambles, bushes, and sun-loving trees like gray birch, sumac, aspen, and juniper. These trees grow fast and die young. They are not good for lumber. Foresters and lumbermen call them weed trees. In a producing forest they are weeds, but in an abandoned field they are really nurse trees, providing a shaded, sheltered environment and improving the soil for future climax trees. So from the stone wall in the woods, we can learn about the history of the area, methods of conservation, and the process of succession.

Succession can also be observed on any abandoned city lot. Despite all the destructive things people do, weeds, brambles, grasses, sumac, Norway maple, and Ailanthus trees do take over. None of these are climax forms, but the succession from bare ground, to grasses and weeds, to woody plants and trees can be observed over several years. Changes in the composition of the vegetation may be observable from the fall to the spring as tree seedlings appear in increasing numbers.

In studying succession, we really are recognizing the way in which one group of organisms changes the environment and in so doing prepares the way for another group to follow.

At one time, my husband and I had a fern garden in which we attempted to grow all the ferns that were native to central Massachusetts. As already indicated, this can be a good study in microhabitats. It can also be a study in interrelationships. (We couldn't grow grape ferns because insects fed on their leaves, and a chipmunk made a tunnel under the microhabitat.) And finally it became a study of, or a battle with, succession.

One of the most interesting ferns of the eastern United States is the climbing fern, which grows in bogs. It had once been common in central Massachusetts, but indiscriminate gathering for sale to florists and draining bogs has exterminated the species in the area. When we found a place where we could purchase it, we prepared a microhabitat by digging a hole a meter deep and 5 meters by 1 meter. We lined it with layers of polyethylene, filled it with black, peaty soil from a nearby bog, and planted sphagnum moss.

When the climbing fern arrived, we planted it in our microhabitat with sundews and other bog plants. It flourished and its beauty gave us much pleasure, but it was obvious that the bog was affording others pleasure, too. Almost daily there were holes, as though an animal had waded through the bog enjoying the moisture. Occasionally we saw a gray squirrel there.

The next spring the bog was a miniature forest. Oak trees, pignut hickory, and wild cherry pushed up through the sphagnum moss. Obviously the squirrel had used the soft bog as a storage place for winter food and then had been unable to retrieve the food when it became ice-encased. Obviously, too, many of the little trees were going to die competing for water, sun, and minerals. But at the same time they would change the bog to dry land as the roots of the living trees took up water and the decaying leaves of the dead trees added solid material. The squirrel, by its activities, had hastened succession. We battled the process by consigning the tree seedlings to the compost heap. In a sense, all gardening is a battle with succession.

While succession is a natural process that occurs relatively slowly and is counterbalanced by geological and meteorological forces, human activities sometimes speed up the process to a highly destructive ratio. The death of the New Jersey salt marshes, the nearly mortal illness of Lake Erie, and the threat to the Florida Everglades are all examples of greatly accelerated succession due to mismanagement of our environment.

At regular intervals there is always someone who writes an article declaring that all is well. It often begins something like this: "Concern about these things indicates lack of knowledge, an alarmist complex. Succession has always gone on. It always will."

Before we settle for this line of reasoning, we need to remember two things. Succession is a natural process like erosion and burning. Like them, it helps make this planet the unique ecosystem which we enjoy. Out of control, it is as threatening as the destructive forces of human-stimulated erosion or the terrifying flames of wildfire leaping through chaparral, forest, or city dwellings.

The second fact may be even more sobering. One definition of succession states that it is a natural process by which organisms make the environment unsuited for their own offspring, with the result that they are ultimately succeeded by other forms. The young of the sun-loving trees cannot survive in the shade cast by their parents. The young of overproductive aquatic algae cannot survive in the lake dried up by the accumulation of the bodies of their parents. Since this process is normally slow, other areas open up and are available and the species does not die out. But where the change is abrupt, organisms become extinct. In speeding up succession in many places, are we, too, creating an ecosystem where our children and grandchildren cannot survive?

RELATED CLASSROOM ACTIVITIES

The study of succession can grow out of a news item or the discussion of a local problem or project.

It can be geography-based. Large parts of Ireland as well as Michigan, Minnesota, and Maine are underlain with accumulations of peat. These lands were once lakes which were gradually filled with plant materials.

It can be geology-based. Often modern succession follows the pattern of ancient geological ages. Algae, the first land plants, prepared the way for higher plants by adding organic material to the soil, by secreting chemicals that affected the rocks, and by holding water in the primitive soils. They were succeeded by more complex plants like mosses and ferns, which also carried on these processes and in addition offered food and shelter for the first land animals. As the bodies of these organisms were added to the rock fragments and dissolved minerals and as the plant roots and animal tunnels aerated the soil, they gradually made way for the flowering plants, grasses, wild flowers, bushes, and trees and for the hosts of animals associated with them.

TEACHER PREPARATION

Succession is a topic that cannot be easily observed in the classroom. But almost any school ground will have evidence of the type of succession which duplicates geological succession. In addition to growths on rock outcroppings and stone walls, steps, cement pavements, asphalt play areas, or the base of the building are exciting places to discover different types of succession.

FIELD TRIP POSSIBILITIES

1. A field trip to locate the succession associated with rocks may begin by observing the growth of algae or of algae and lichens on a stone outcropping or wall or by examining moss in a pavement crack. Frequently, some grass, an Ailanthus tree, or a flowering plant may be found growing in these same areas, where some organic material has been added to the harsh mineral soil of broken rock or cement. Regardless of the starting point of the lesson, succession is an exciting concept anywhere and particularly on a "paved desert" school site.

2. Many school grounds have a neglected corner, untended by the custodian and untrodden by the school's baseball team. Such a spot offers a class a good outdoor laboratory to study succession. The wide distribution of seeds is one reason why succession is so universal. Year after year an area may be bombarded with seeds which die for lack of proper conditions but, finally, when the conditions change, some of the seeds germinate and a new chapter is written.

This was true when fireweed clothed the bomb sites of London in beauty. Fireweed seeds had always blown into London and died on the streets, buildings, and pavements but when these were demolished and the soil was exposed, they took root and started healing the soil and making way for more demanding plants.

Can your class discover the source of the plants in your unkept corner? Can they find other places where seeds that have accumulated could take root if conditions changed?

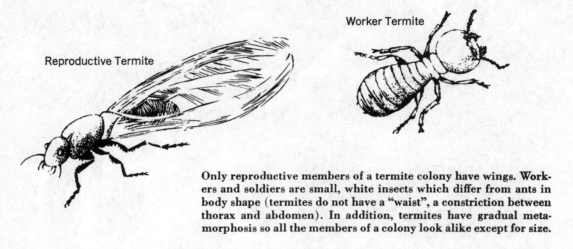

Reproductive Termite

Worker Termite

Only reproductive members of a termite colony have wings. Workers and soldiers are small, white insects which differ from ants in body shape (termites do not have a "waist", a constriction between thorax and abdomen). In addition, termites have gradual metamorphosis so all the members of a colony look alike except for size.

3. Often a log tells an interesting story of animal and plant succession. Some invertebrates, like carpenter ants, termites, and beetle larvae eat only dry, tough wood fibers. Others, like millipedes and sow bugs, feed on the softened decaying plant material that has soaked up moisture in the tunnels made by the first group of animals. Bacteria and fungus plants also obtain food from the rotting log. Gradually they change it to a soil-like substance which supports moss, earthworms, and finally ferns, grasses, weeds, and even young trees. Sometimes one part of the log will still support the early animals while another section, where decomposition is more advanced, will have their successors. Sometimes adjacent logs may represent different stages in succession.

With magnets in hand, even small children can make extensive (and sometimes amazing) discoveries about building materials in the school area. These discoveries in turn may raise new questions for further study indoors and for later trips outdoors.

Physical Science ~~~~~~~~~~~~~~~~~

The study of science in elementary grades is frequently divided into biological, physical, and Earth science concepts. By definition, biological science deals with living things. Earth science deals with the geology and weather of this planet, as well as the relationships of the Earth to the other parts of the universe. Physical science is the study of the natural laws and processes of the nonliving world.

Obviously, Earth science is really a part of physical science. Furthermore, physical science processes affect and are affected by both biological organisms and the forces studied in Earth science. For instance, sounds are waves set in motion, usually in air, by a vibrating object (a physical science concept); recognition of sounds depends on ears or other biological receptors. In addition, sounds are produced either by people and other animals (biological organisms) or by rain, wind, and other weather phenomena or by geological forces like ocean waves, volcanoes, and earthquakes (all Earth science topics).

Another example of these interrelationships is light and heat. The physical scientist defines light and heat in the abstract, but the effect and intensity of both are determined by geological components, and both strongly influence and are influenced by biological organisms.

This blending of the sciences is apparent whenever one does environmental studies. In fact, environmental studies will frequently go beyond the natural sciences to include psychological and social science aspects. For this reason, school ground experiences in physical science topics are excellent because they move abstract laws from the laboratory and test tube into the reality of life.

As youngsters gain experience with physical science concepts, they will begin to discover that in all instances we could not get along without these natural forces, but by the same token, every one of these forces sometimes creates problems. This is equally true of gravity, friction, heat, light, sound, electricity, physical change, chemical change, and all sources of energy. Frequently, problems concerning these natural forces are created by human activities.

RELATED CLASSROOM ACTIVITIES

Many physical science concepts that were known only to scientists have come into everyday conversation due to television reports on space exploration. Gravity is much better understood when we see what happens when we are out-side of its pull or when its pull is much reduced as in space or on the moon. Friction, the atmosphere, and radiation are other topics that have taken on new meanings as a result of space exploration.

Whether a physical science topic is introduced from a news item, from a child's question, or by the teacher, it can probably be enhanced and brought down to Earth by saying, "Let's go outside and see. . . ."

TEACHER PREPARATION

Since the topics studied in physical science have universal applications, they can be studied, observed, and sometimes experimented with on any school ground. Teacher preparation will involve selecting the best approach for the age group, the area, and the topic at hand, and deciding on class organization for maximum participation.

FIELD TRIP POSSIBILITIES

1. The school ground is often a better place than the classroom to experience and observe the pull of gravity simply because jumping, tossing things into the air, and dropping things is less disruptive and safer outdoors than indoors. After children have consciously experienced the pull of gravity, discuss its influence on our lives. The list can be almost limitless. For instance, gravity keeps us from floating off into space, brings the tossed ball back to Earth, and holds onto the air that makes Earth such a special place.

2. The pile of broken rock material at the foot of any rock formation (such as a brick or stone building, a pavement curbing, a cement step, a stone wall, or a rock outcropping) provides a quiet way to observe the influence of gravity. These small accumulations of rock fragments are miniature editions of the big piles of broken rock, called talus slopes by geologists, that can be found at the foot of any cliff or mountain. Regardless of size, gravity has pulled the loosened material down to the bottom of the slope. Sometimes where cracks occurred in an over-hanging section, gravity has helped with the disintegration of the rock by its pull on the unsupported fragments. This may frequently be observed on a stone or cement windowsill or on building trim which extends beyond the building surface.

3. Snow may sometimes be observed as it is pulled off a slanting roof by gravity in the same way as snow, ice, and rocks are at times torn loose from a mountainside in the form of an avalanche. Modern public buildings are con-

structed to avoid the danger of a big snow slide, but homes, garages, and other buildings with slanting roofs may be observed from the school area. Can children see that in planning a building in an area of heavy snow, the architect must consider the effect of gravity? The downward pull (weight) of a heavy snowfall can cause a flat roof to collapse unless it is built with special reinforcements. Sometimes homeowners must shovel their roofs.

If you live in a snow belt, what provisions are made in your school and community to protect people from the pull of gravity on snow on roofs?

4. Even though the pull of gravity is equal on a fragment of rock that has broken loose on an overhang and an identical fragment at the top of a talus slope, the former will get to the ground faster, all other things being equal, because air offers less resistance than the jagged pieces of rock that make up a talus slope. This resistance that occurs when one body moves across another is called friction.

Friction is one of the forces that helps us overcome gravity, but too much friction requires the outlay of a great deal of energy. Often in machinery, we try to reduce friction but not eliminate it. In winter, observe cars moving on snow or ice. If tires are smooth and friction has been reduced too much, what happens, or doesn't happen? What do snow tires and chains do? Put sand or ashes on one section of snow or ice. Compare walking on this section of the school ground with walking on ice or packed snow. Compare walking on ice with smooth rubber soles, heavily ridged soles, and metal ice creepers.

5. Make a survey of your school site to find spots where you can compare areas with varying amounts of friction, for example: a metal flagpole and a tree trunk, a grassed area and a clay area after a rain, a sandy hill and a grassed or rocky one, a smooth concrete sidewalk and a sidewalk with a rough texture.

6. Examine your baseball diamond. Has it been cared for so that weeds and other objects that would slow the players down with too much friction have been eliminated while enough friction has been provided to make it possible to run the bases? Could anything be done to improve your school's diamond by increasing or decreasing friction?

7. The classroom is the best place to discover the properties of magnets and the things which magnets will and will not attract. After experiments with magnets have been done and some conclusions summarized, go out on the school ground and make a search for all the places in which iron and steel have been used in and around the building.

8. The only way to really understand how a compass serves as a guide in finding directions is to use one. Let one group of children lay a trail or make a treasure map and another follow it. The map would say something like: "Go out the front door. Go ten steps (paces, feet, yards, meters, or other agreed-upon unit of measurement) due north. Turn east. Go five steps, turn southeast 15 steps. You should now be under a tree. Turn due west. . . ." This kind of experience could be made possible by two teachers working together or one teacher with the help of parents or teacher aides, or a group of children might prepare the map after school.

Physical and
Chemical Change

The Earth and everything upon it is composed of 92 naturally occurring elements. Each element is built of electrical units arranged in such a way that they cannot be broken down (except by atomic fission). Each element has certain unchanging characteristics including behavior, weight, appearance, and the arrangement of electrical charges.

Of the 92 elements, nine make up 95 percent of the Earth's surface. These nine are: oxygen, silicon, aluminum, iron, calcium, sodium, potassium, magnesium, and hydrogen. Except for oxygen and hydrogen (invisible gases which occur as free elements in the atmosphere), all of these elements occur naturally only as compounds. In addition, carbon, chlorine, and sulfur are common in many compounds. Even though carbon occurs in relatively small amounts, it is tremendously important, since it is the main component of the organic compounds of which all life is composed.

A compound is a substance made up of atoms of two or more elements combined on the basis of their electrical charges. This is called chemical bonding. A molecule of a compound cannot be broken down into its atoms by physical means.

For instance, most iron occurs in the form of iron oxides. Hematite is one kind of iron oxide. One molecule of this compound is composed of two atoms of iron and three of oxygen. It is a nonmetallic rock that makes a maroon streak when it is pulled over a piece of unglazed porcelain. It can be hammered into bits, but every little piece is still hematite. A magnet can be pulled over it but nothing happens; the compound does not respond to magnetic pull. The iron cannot separate from the oxygen. To break down the compound, hematite is heated to a high temperature with limestone, a compound composed of calcium, carbon and oxygen, and with coke, which is 90 percent carbon. When this happens, the oxygen of the hematite combines with the carbon to make a new compound, carbon dioxide, and leaves pure iron behind. Just as the iron does not resemble the hematite, so carbon dioxide does not behave at all like oxygen, nor does it suggest in any way that it is partly composed of the solid black substance we call carbon.

Green plants take the compound carbon dioxide and use the Sun's energy to combine it with water, a compound composed of hydrogen and oxygen, to make sugar and release oxygen.

The oxygen thus released may combine with iron to form an iron oxide. Or it may enter the body of an animal and be used to oxidize food and become part of a molecule of carbon dioxide or water, or it may be used by a burning candle, or in numberless other ways. The constant recycling and reorganization of the molecules of the Earth's elements help make life possible.

There is no predicting where the oxygen we breathe at any given moment was last in either its chemical combination or geography. One thing is certain, however, in all chemical change: Energy transfer is involved. The energy may be necessary to bring about the change, or it may be released as the change occurs. Thus, the Sun's energy in the form of light is necessary for photosynthesis. When the food produced through photosynthesis is oxidized by either plants or animals, the energy that was stored is released for growth, reproduction, movement, and other bodily activities.

When this release is rapid, we feel the heat as in a burning candle or even in our own bodies. When it is slow, as in rust, the heat is dissipated and can only be detected with sensitive instruments.

RELATED CLASSROOM ACTIVITIES

Lumps of sugar can be used to experiment with physical and chemical changes. The sugar can be broken into bits, dissolved in water, and recovered by evaporation. These are all physical changes. But if sugar is burned, it produces carbon and releases water and carbon dioxide. This is a chemical change. In the first series of activities, taste, appearance, and weight will all remain the same. In the second, they will all change. Another method of producing a chemical change in sugar would be to dissolve it in warm water and add yeast. As the yeast uses the sugar for food, it produces bubbles of carbon dioxide and a weak solution of alcohol. The bubbles can easily be observed, as can the change in both taste and smell.

Wax can also be used in simple experiments to show physical and chemical change. The wax can be chopped up, melted, shaped, and reshaped but it remains wax. But if a wick is added and energy in the form of a flame is applied, the wax combines chemically with oxygen, and molecules of carbon dioxide and water are formed. Carbon which fails to combine with oxygen becomes apparent as black soot. Thus, the wax which is a compound of carbon and hydrogen becomes water, carbon dioxide, and carbon when it undergoes a chemical change.

To observe the change from wax to new compounds, a jar may be inverted over the burning candle. If the jar is cold, small droplets of water will be deposited on its sides.

As the oxygen in the jar is used, the flame dies out. Carbon may be deposited above the flame or appear as particles of smoke. The carbon dioxide formed in the burning will accumulate in the jar and can be identified by pouring lime water into the jar. Lime water is calcium oxide dissolved in water. When carbon dioxide contacts calcium oxide, the colorless gas combines with the colorless liquid to form a white deposit, calcium carbonate.

Once children have some experiences with physical and chemical changes, they will be able to discover other examples in their homes and their daily lives.

Photosynthesis, digestion, and oxidation of food are only a few of the processes that involve chemical changes in living things. Perhaps the most easily observed physical and chemical changes can be demonstrated with a box of soda crackers. Break a cracker into small bits, a physical change. A small piece of cracker tastes like, generally appears like, and is like a whole cracker. But if the cracker is thoroughly mixed with saliva and held in the mouth for several minutes, a chemical change takes place as starch is converted to sugar. This change in flavor can readily be recognized.

TEACHER PREPARATION

A preliminary survey of rock outcroppings or rock (stone, brick, or concrete) structures will reveal teaching possibilities on the school grounds for either physical or chemical changes or both. Teacher preparation need not go any further than locating a teaching site to which the class can go to discover what kind or kinds of change are occurring.

Discovering copper compounds may be as simple as stopping to scan the building for green staining from the roof or drain pipes.

A search for iron oxides is one field trip which can be made with only organizational preplanning. Iron is present on every school ground in things like fire escapes, fences, grills, and nails. The various examples of iron oxide can best be discovered by an interested class of youngsters turned loose on the grounds, after they have been shown a sample of the iron oxide compound we call rust.

FIELD TRIP POSSIBILITIES

1. Have the children find evidence of physical change in the form of rust on fire escapes, fences, grills, and nails. The teacher should scrape some rust off each object. (Care should be taken when dealing with sharp objects.) Back in the classroom, compare the rust to iron filings in color, response to a magnet, strength, and any other tests the children can devise.

2. Look for methods which have been used to prevent oxygen from contacting the iron and combining with it. Are they all equally effective? Can children find evidence of chemical or physical changes taking place in the substance used to protect iron and steel, like oxidation of paint (chemical), peeling of substances like chrome plating and paint (physical change)?

3. If rock outcroppings or stone walls contain black iron compounds like biotite mica or hornblende, they may be streaked with iron oxide as the large molecules of these minerals break down into several substances composed of smaller molecules. This, too, is a chemical change, and the properties of the new brown compound should be compared to the properties of the original minerals.

4. In areas where water freezes and thaws throughout the winter, examine sidewalks, walls, and rocks for physical change; that is, for evidence that big rocks are breaking into small pieces due to the expansion of freezing water which collects in the cracks. Geologically this change is called disintegration. It is most important in temperate zones where temperature fluctuates. It does not exist where there is no freezing, nor for that matter where there is no thawing. Layered rocks or rocks made of interlocking crystals are particularly susceptible to physical breakdown.

5. Copper exposed to air combines with carbon dioxide to form a green deposit of copper carbonate. This green deposit is often noticeable as staining on a building. If it can be reached, a sample of it may be collected and taken to the classroom for comparison with copper.

6. Insects working on a log may be responsible for either physical or chemical change. By gnawing tunnels and galleries, carpenter ants chop up the wood and toss it aside as sawdust (physical change). Termites and many beetle grubs eat the wood. When it leaves their bodies as fecal material and gases, it has undergone a series of chemical changes and has become a number of new compounds.

Sound

A study of sound can deal with very simple ideas or highly complex mathematical formulas.

However, as soon as the pre-primary stage of identifying the sounds that different organisms make or the different sounds that one organism can make has been completed, the study of sound always involves vibrations.

Children can feel their own "voice box" vibrating by putting their hand lightly on their neck just under their chin while they sing, talk, hum, or growl. Not only will they feel the vibrations but they will be able to feel that the vibrations are not all alike. Low-pitched tones come from slow vibrations and are much easier to feel.

In sound production, the vibrations set in motion by an organism or object always travel through other media. Unlike radio waves and light waves, sound waves cannot travel in a vacuum. Men on the moon communicate by radio waves. Without air, speech would not produce sound waves. No matter how astronauts' voice boxes vibrated, sound waves on the moon would have to travel through the soil, and both speaker and listener would have to lie on the lunar dust and communicate with a series of previously agreed-upon thumps and scratches.

So far no studies have been made of sound conduction in lunar dust, but from Earth studies we know that sound usually travels fastest through dense materials. For instance, it travels at approximately one-third of a kilometer per second in air; faster when the air is heated, slower when it is cold. (This is unusual because air is less dense when warm.) It travels roughly three kilometers per second in a pine log, five kilometers per second in glass or steel, almost 1.5 kilometers per second in water, and 1.3 kilometers per second in hydrogen (another time when sound travels unexpectedly rapidly).

The fact that we have determined the speed of sound in different substances makes sound a useful measuring device.

Bats have been traveling in the dark for eons by using reflected sound. The high-pitched sound that sometimes can be heard as bats travel overhead is the lowest of their sounds. Most bat sounds are so high-pitched that human ears are unable to catch and interpret their vibrations. As the bats squeak overhead, the sound waves hit objects and bounce back (echoes) to the bat's sensitive ears. By this process, bats fly safely without crashing into anything and also locate flying insects (the only source of food for U.S. bats).

Today this kind of measuring technique, called sonar (*sound navigation ranging*), is used for determining the depth of bodies of water and for locating mineral deposits. The relative speed of sound and light can also be used to locate things. People often locate lightning in this way during thunderstorms. Although the light flash and thunder clap occur simultaneously, unless you are at the spot where the lightning strikes, the thunder seems to follow the flash. This is because light, traveling at almost 300,000 kilometers per second, is seen instantly for all practical purposes, but sound, traveling through air at 0.3 kilometers per second, takes time to arrive. So measuring the time between the lightning flash and the thunder clap will give an approximate distance for the place where the lightning struck. Thus, two seconds would be 0.6 kilometers, five seconds for 1.5 kilometers, ten seconds for 3.2 kilometers. Sometimes lightning seems to light up the horizon but no thunder is heard. When this happens, the storm is so far away that the sound does not reach us because sound diminishes in volume very quickly with distance.

RELATED CLASSROOM ACTIVITIES

Any study of sound should involve listening, not to the teacher talking, but to a wide variety of sounds produced in a variety of ways. Children can bring sound-makers to school, or they can make their own from all kinds of easily obtained materials.

Concerts may be produced, ranging from rhythm bands for younger students to orchestras and solo numbers produced on instruments which have been tuned for pitch by older children. From making instruments, the youngsters will be able to discover the laws of sound production and sound waves.

Other experiments and observations can be made on topics like sound traveling through various media, sound reflection, directing sound toward one spot (a megaphone, a rabbit's ears).

A study of sound can grow out of a study of music. In this case, a music supervisor and a science consultant may serve as resource people for a classroom teacher and her pupils.

Today hearing aids are becoming common. Much can be learned about sound and sound reception by studying the different types.

TEACHER PREPARATION

A few school ground field trips for sound require previous checking. If you are going to tape the song of a cricket or the sound of a basswood tree with thousands of humming bees collecting nectar, you should be reasonably certain they will be there. But some sound experiences simply require space and previous planning concerning the route to be followed and the time to go, so that the class's sound experiences will not be drowned out by the sounds of a school baseball game.

FIELD TRIP POSSIBILITIES

1. Take a let's-depend-entirely-on-our-ears field trip to the school ground. How many things can the class identify? Why do they hear more things with their eyes closed than when they are open? Besides naming the sound makers, can they locate the direction of sound? Tell if the sound maker is moving? Tell if the sound is pleasant or unpleasant?

2. Blind people depend on sound a great deal. Help children develop a sensitivity to some of the problems of the blind by keeping their eyes closed for a period of time. Can they hear people approaching them if they are on asphalt or concrete? On grass? Can they tell if someone is beside them? Has passed them? Can they identify their friends? What can they tell about traffic? Can they walk directly to someone who calls them? To a ringing bell? What things would they miss if their ears had to do the job that their eyes help with? What things would they enjoy more? Conversely, what would they miss if they had no hearing?

3. Noise pollution is a twentieth-century problem. What are the effects of excess noise? Have children try to perform some exercise in complete silence, then repeat the exercise in loud noise. How does this affect concentration? What other effects can be observed? Listen in on your site. If sound is a pollutant, is there anything that can be done about the source of the noise? (Yelling teachers and children can learn to drop their voices, cars can be required to have mufflers, broken highways can be repaired.) Is there anything that can be done to buffer the sound? (Carpet deadens the sound of feet, a garden area or lawn separating the play area from the building lessens the impact of voices, a city can change its traffic patterns by closing off streets during certain hours, rubbish collection can be scheduled for before school or after school hours.) Let children stop, listen, and then look for reasonable solutions.

4. If a portable tape recorder is available, record the same sounds out in the open, under a tree, under bushes, and in the entrance of the building. Where are traffic noises loudest? Softest? Is this the same for airplane noises? What other kinds of sounds can you compare? What conclusions can the class draw from their evidence?

5. Divide the class in half, giving half the class a noise maker—a drum, a ratchet, a bell—or just have them use their own voices. Have this group stand on one side of the building. Have the other half of the class directly opposite them on the other side of the building. (The school building should be between the two groups.) Can the second half hear the first? If the two groups stay on each side but move to the front of the building so the sound will have to turn the corner, can the first group be heard? If both groups move beyond the building but remain at the same distance from each other, can the noise makers be heard? Reverse the experiment, letting the noise makers become listeners and the listeners make the noise. What conclusions can the class make about the way in which sounds travel? A tape recorder could be used in this study.

6. If planes fly over your school ground, it is easy to observe the difference in speed of sound and light. Have half the class point to the place the sound comes from. Have the other half point to the plane. For all practical purposes, light hits

your eye at the same moment that it hits the plane, therefore you see the plane where it is. Sound traveling at roughly 0.3 kilometers per second takes 10 to 15 seconds to get from the plane to the Earth, if the plane is flying at 3,000 or 4,500 meters. In that time, the plane traveling at several hundred kilometers per hour has moved forward 15 to 30 kilometers. So the sound as heard on Earth always trails behind the plane. How far it trails behind will be dependent on the plane's altitude and flight speed.

7. If you can get far enough away from your building, you may be able to hear sound bounce off its surface (an echo). Notice that as you get closer to the building, the time between the noise and the echo grows shorter, until you are finally unable to hear the echo.

With older children, use a stop watch to measure the time between the sound and the echo. Or count by saying one second, two seconds, three seconds, four seconds. If sound travels at 0.3 kilometers per second, can they estimate the distance to the building? Measure the distance. Minor differences will be due to the fact that 0.3 kilometers is an approximate figure which varies slightly with temperature, humidity, and altitude. A major difference may require retesting, remeasuring, and rethinking. Does their answer represent the distance the sound traveled from them to the building and back to them or does it represent the distance from them to the building?

8. To determine the direction in which sound waves travel, have one child stand in the center of a circle with a noise maker such as a bell, rattle, or ratchet, which can be held overhead. Arrange the children in a circle around him. Can they all hear the noise? Have the children each step back two steps making a larger circle. Can they still all hear the noise? Repeat stepping back two more steps. If one child cannot hear, have him exchange places with someone who did hear. Is it the spot or the child's hearing? Continue testing with larger circles. Have the children draw a conclusion about the direction in which sound travels. Did they make any observation about what happens to the volume of sound as the distance from the source increases?

9. Indians used to put their ears against the ground to listen for approaching animals and people. Have one group of children listen with their eyes closed and their ears to the ground while the other group approaches. Can they tell what is happening? Can they tell anything about numbers? Speed? Reverse groups and repeat the activity.

Heat and Light

Heat and light are the two best-known forms of energy. Scientists believe that, like radio waves, ultraviolet rays, and X rays, heat and light are electromagnetic in origin. All of these are energy sources which travel in a straight line at a speed of 300,000 kilometers per second.

Over the years, we have gained much knowledge about the ways in which these rays behave and the ways to use and produce these fantastic bundles of energy, but we still have not discovered the secrets of their origin or composition. (Newton was one of the first scientists to propound theories on this topic and Einstein was one of the most recent.)

Except for atomic heat and light, which come from fusion (for example, in the H-bomb) or fission (for example, in the A-bomb and in nuclear reactors used to produce electricity), all elements of heat and light can be traced to the Sun.

Since the Sun's energy, to the best of our knowledge, is totally atomic in origin, fusion and fission duplicate on Earth the source of solar energy. It is interesting to note that helium, a product of the fusion of hydrogen atoms, was discovered on the Sun before it was recognized on Earth, and was named after the Greek word *helios*, the Sun.

That does not mean that we can shrug off the by-products of atomic energy as natural and a matter of no concern. For a long time, we have known that one reason for the success of life on this planet is our distance from the Sun: near enough to benefit from the energy of the electromagnetic rays of heat and light, far enough away to receive these and the other cosmic rays in a greatly reduced quantity.

While it is the energy from the Sun that makes life possible on Earth, living things are also responsible for the storage and recycling of much of this energy. Green plants combine carbon dioxide and water to produce food using the energy from light. The energy thus stored may later be obtained by oxidation of this food by the plant itself, or by an animal that eats the plant, or by the burning of the wood, dead leaves, corn cob, or other plant parts containing the stored food. The heat may be slowly released by the process of decay. Or the plant may die in a moist acid habitat and slowly become peat, thus storing the energy for future use. Coal and oil are both heat and light sources derived from long geological processes on plants or on animals which fed on plants.

Even water power is produced by the Sun's energy through the evaporation of water, followed by rainfall on uplands, and the subsequent downward rushing of streams.

Therefore, it is not surprising that many early peoples living in close harmony with the Earth worshiped the Sun. Without knowledge of corpuscles, waves, photons, or other scientific concepts, they were sensitive to the fact that heat and light were central to life and came to them from the Sun.

Solar heat travels in the form of infrared rays. We believe that when these rays strike a solid object, they are absorbed and generate heat by setting in motion the electrical particles in the molecules of the object.

Outer space is both cold and dark, in part because there are no objects to intercept or be influenced by the infrared rays and the light rays streaming from the Sun.

Frequently, we describe materials by the way in which they affect electromagnetic rays. Transparent objects like air, water, and glass permit the passage of infrared and light rays. As the rays move from one transparent substance to another transparent substance of a different density, they are bent. We call this

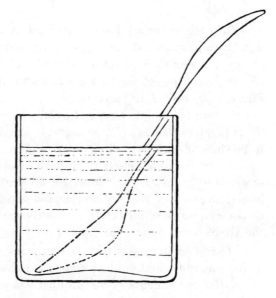

Refraction Demonstration

Refraction, the bending of light as it passes through transparent objects of different densities, can be readily observed by viewing a spoon or other object through water and a glass container and comparing the relative amount of "bending."

process refraction. It can be observed by putting a spoon in a container of water and noticing the apparent bends it acquires when it leaves the water surface as well as the jar surface.

When light and infrared rays hit a shiny, solid substance, they bounce off, in a process called reflection. Mirrors are well-known reflectors; so are headlight reflectors, and aluminum foil reflectors under stove burners.

An opaque, solid substance creates a resistance to the infrared rays and converts them into heat energy. Once heat energy has been generated, it passes from molecule to molecule by the process of conduction. Some substances are good conductors of heat; others are poor ones. Cooking pans are made of substances that conduct heat well. For practical purposes, their handles should be made of materials that conduct heat poorly. When they aren't, the hot pan must be handled with a pot holder which is made of a poor conductor. Otherwise the heat will be conducted to the cook's hands with serious results.

When a substance is heated, it expands. This means that a given volume of a cold substance weighs more than an equal volume of the same substance at a higher temperature. In liquids and gases which have no fixed shape, the expanded warm molecules move upward, while the more dense cold ones sink downward. This pattern of motion is called convection. It is extremely important in the distribution of heat.

Before the astronauts who landed on the moon left their spacecraft, they put on specially constructed space suits that could withstand temperatures ranging from 120°C above zero to 120°C below zero.

Why should there be such a tremendous range of temperature on the relatively nearby moon which, like the Earth derives its energy from the Sun? Why should it be so dramatically different from the Earth?

First, the upper layer of air above the Earth and the clouds reflect a great many of the rays which radiate (travel through space) from the Sun into space. The moon has no such protection. Second, the moon turns on its axis once in approximately 28 days, so any given spot is bombarded by the Sun's energy for a period of roughly 14 days as opposed to the Earth's half-day exposures. Third, without air or water, no convection currents distribute the heat, so the side facing the Sun has no influence on the unwarmed surface which is turned away from the Sun and vice versa.

RELATED CLASSROOM ACTIVITIES

Children can have many experiences with heat and light in the classroom. They can devise experiments to find answers to questions like: "How does the whole room get heated when the radiator is only on one side?" "How can we protect our aquarium from too much sun?" "Is there anything we can do to cool our room on a hot day?" "How does an air conditioner work?" "Where does the hot air in a refrigerator go?" They should also have an opportunity to experiment with a wide variety of substances in relationship to light and heat and observe processes like refraction, reflection, absorption, conduction, and convection. To compare the speed of heat traveling by radiation to the speed of conduction or convection, let a child put one hand lightly on a light bulb and pull the switch with the other. Heat (and light) travels through the vacuum of the bulb by radiation. Now put a pyrex beaker filled with water on a burner. Put a finger in the water and turn on the burner. Does the water temperature change immediately? Does the temperature of the top of the beaker change immediately? Heat is conducted through the glass and the water.

Can you feel convection currents in these situations? Let the children devise techniques for showing that the heated air is rising.

TEACHER PREPARATION

All school sites have a variety of spots that show the effect of different temperatures caused by shadows, the reflection and absorption of different materials, vegetation, and differing exposures to sunlight. Preparation for a field trip should consist of becoming aware of these spots and choosing the ones best suited for the topic under discussion. If thermometers are used in temperature studies, it is important to be sure they are all synchronized. This can be done by putting them all on a desk top and checking their readings after 10 minutes.

FIELD TRIP POSSIBILITIES

1. Objects both reflect and absorb heat and light in varying degrees. You see the side of a building because of the light reflected from it. At the same time, it is warmed as it absorbs infrared rays and converts them to heat. Some of this heat is transferred to the air by conduction; some is reflected from the surface and is absorbed by the soil and the air. Often there may be a several-day or several-week difference between the blooming time of plants growing against the wall of a building and plants growing 0.3- to 0.6-meters away from it. This is particularly true with early-blooming bulbs like crocuses and daffodils. If you have a site of this type, try to discover all the heat and light stories involved.

2. A remarkable variation in temperature can be discovered at any site. Even air temperatures will vary drastically if taken 30 centimeters above the grass, on an asphalt parking lot, on a concrete walk in direct sunlight, or 30 centimeters above the same areas in the shade. Compare temperature readings taken on the surface of each of these areas; taken 1 meter above the surface. Make graphs to record your findings. Can you draw any conclusions about the absorption and conduction of these materials? About convection currents?

Try doing this study in different seasons. Or compare the temperature of grassed, concrete, asphalt, and bare soil areas early in the morning, at noon, and in late afternoon. If possible, have someone make an evening study in the same areas to complete the statistics. You will discover that not only are there variations in the ability of substances to convert infrared rays to heat and in the ability to conduct heat, but also in the ability to retain heat. An area dominated by substances that heat and cool quickly will have a different climate from one dominated by substances which change temperature slowly.

3. Snow and ice provide an easy way to make temperature studies. Find places where a dark object, which absorbs heat well, has melted into the snow. Look for the effect of objects on the snow, of tree trunks, of buildings. Put objects of different colors on the snow and watch their effect. Have children draw conclusions about light colors and dark colors in connection with reflection and absorption.

4. Look at house roofs for heat stories. Can you see some where the snow has melted because of the way the roof faces the sun? Are there any roofs visible from the school ground where the snow is melted around a chimney? Can children find other places where the snow or lack of it tells about heat escaping from a house?

5. Icicles or a section of ice in an area of snow tell of a warm period followed by a cold one. This may be a warm roof that melts snow even though the air is freezing so that the water dripping from the roof forms icicles, or it may be a warm spot where warm sunlight that melts snow for a few hours is followed by freezing cold which forms ice.

6. In the spring, small patches of snow in sheltered spots tell of cold areas. Map these spots and try to determine why they are colder than the surrounding areas.

7. Even solids expand and contract in response to temperature changes. A big solid would buckle if it got too hot in the middle of the summer. Make an inventory of things that have been done to prevent buckling, like sidewalk cracks, spaces between pieces of metal on a fire escape, and gratings.

8. If you have a school parking lot where cars are parked at approximately the same time of day facing in the same direction, compare the temperature of different-colored cars by using hands to judge warmth. Be sure color is the only variable. Compare the temperature of several parts of one car. If there are differences between the hood, roof, side and the back above the exhaust, what factors might be significant? If there are differences, what part of the car do the children think they should check for the best comparison of the absorption of heat by different colors? What other things might be compared?

9. Using hands for temperature receptors, do a study of the temperature of two sides of a tree, the end of a metal railing or fence in the sun and the end of the same railing or fence in the shade, the riser and the tread of a concrete step, and other substances like asphalt, concrete, and grass. Which was hottest? Coolest? Record your findings. Do the same thing on a cloudy day. A winter day. Use records of the three trips to interpret your findings.

10. Smoke is a combination of warm gases and unburned carbon. Watch it come out of a chimney. Which direction does it move? Does wind change its direction? How far up does it go before it levels off, changes direction, and disappears? Wind is like a convection current inasmuch as heavier cold air pushes warm air around. Watch the smoke coming from a chimney for several days. Does the wind always come from the same direction or different directions? If wind comes from the same direction for a period of weeks or months, can the children find out why?

11. For several days or weeks, check the temperature under a tree and in a treeless spot that duplicates the place where the tree is growing in all respects. From your results, decide whether it might be a good idea to consider planting trees for their effect on temperature. If possible, check an evergreen tree and a broad-leafed tree. Study both trees in warm and cold seasons. Do they both have the same influence on temperature in summer, fall, winter, and spring? If you were only considering temperature, might you choose a different tree than the one you might plant if you considered growing conditions as well?

Exposed bedrock offers many learning opportunities. Here an intrusion weathers faster than the mica schist and provides a place for plants to take root. As soil forms, disintegration and decomposition rates speed up, providing a mini-geological laboratory to discover the relationships of water, temperature, organic acids, and the growth of roots.

Earth Science

If we look closely at the moon, we will begin to develop a greater appreciation for the Earth. Without air, water, and soil, the time we can spend on the moon is limited by the amount of water, oxygen, and food that can be carried there.

Air and water both escaped from the moon long ago because of its weak pull of gravity. Without air and water, plants cannot grow, nor, for that matter, can rocks become soil.

The surface of the Earth is composed of three interwoven and interacting units: the lithosphere (rocks and soil), the atmosphere (air), and the hydrosphere (water).

Soil is the part of the lithosphere which makes life on land possible. It is a mixture of rock fragments that have been broken by the physical action of ice, heat and cold, plant roots (disintegration), and by the chemical action of acids formed when water unites with carbon dioxide and organic substances (decomposition). Mixed with these small pieces of rock material are parts of dead plants and animals, as well as microscopic living plants and animals like soil bacteria, fungi, and protozoa. Distributed in varying amounts throughout this complex soil mixture are air and water.

The atmosphere (air) is also a complex mixture. Basically it is composed of nitrogen (roughly 78 percent) and oxygen (almost 21 percent). A number of insignificant inert gases (argon, helium, neon, and hydrogen) make up about 1 percent. Two other essential compounds, carbon dioxide and water, complete the composition of this mixture. The amount of carbon dioxide and water present in the air varies with the plant and animal life and other activities of the area. Both carbon dioxide and water are products of the oxidation of food, both are products of combustion, and both are used by plants in synthesizing food. Temperature and climatic factors also influence the amount of water vapor in the air.

The relative amounts of the various components of air are important. Many organisms die when air contains too little water vapor or too much. Lowering the percentage of oxygen in the air makes respiration difficult or even impossible, but a large increase in the percentage of oxygen creates biological problems as well as insurmountable fire problems. This can be demonstrated by putting iron filings in pure oxygen, lighting them with a match, and watching them burn.

From time to time, other substances in varying amounts are added to the atmosphere. This has always occurred naturally. When a flash of lightning passes through the air, oxygen atoms combine forming ozone (O_3), and nitrogen and oxygen atoms combine forming nitrogen compounds. Methane gas rises from a swamp. Sulfur dioxide, hydrogen sulfide, and other gases are produced in the intestines of animals, by the decay of protoplasm and other biological processes, and by volcanic action. Bacteria, viruses, insects and other tiny creatures, pollen, and larger particles are picked up by moving air and become a part of the atmosphere for longer or shorter periods of time.

Many of these are classed as pollutants. But again, this depends on the relative amounts. Small amounts of dust in the air are essential for precipitation. A piece of dust forms the nucleus of every raindrop. Small amounts of ozone are dissipated in the air and finally become stable oxygen molecules (O_2). Nitrogen compounds are essential to plant growth and the formation of proto-plasm. They quickly settle out of the air and into the soil. The distribution by the air of pollen from many trees, most grains, and other grasses is essential for our continued food and oxygen supply. Even poisonous substances like methane gas generally occur naturally in such small amounts that they are quickly diluted in the great ocean of air.

In recent years, however, human activity has added such huge quantities of dust, ozone, carbon monoxide, sulfur dioxide, and other substances to the atmo-sphere that natural absorption and dilution have not been able to function in many areas. In these areas, the result has been an extremely unpleasant environ-ment, the destruction of plant and animal life, and even serious illness and death of people and animals.

The hydrosphere makes up the third part of planet Earth's surface. Although water is a compound composed of hydrogen and oxygen, the hydrosphere is also a mixture. When we speak of pure water, we are describing an extremely rare phenomenon. Rain starts on dust particles, picks up substances as it falls through the air, and carries soil particles as it moves over the soil. Water is often de-scribed as the universal solvent. It not only dissolves many things itself but it forms weak acids when it combines with carbon dioxide and other substances, and these often speed the process of decomposition.

Often the bedrock of an area strongly influences the composition of water. We describe water that contains dissolved limestone as hard. Mineral springs and sulfur springs obtain their mineral content from the rocks below them.

Water from the ocean is indicative of the great collecting and holding power of water. Over billions of years, materials from the lithosphere have been dissolved and eventually deposited in the world's oceans.

Humans have always used water for cleansing and for the removal of wastes. When there were fewer people and waste products were largely organic, this did not pose many problems. Today, however, the organic wastes of a vast population and the tremendous variety and quantity of waste from industry cannot be absorbed or adequately diluted by the hydrosphere, with the result that life in the ocean as well as in fresh water is threatened.

The lithosphere, atmosphere, and hydrosphere are literally three spheres enclosing the inner core of the Earth. At some places, like lakes and oceans, the lithosphere dips down under the hydrosphere; at other it is the hydrosphere which is submerged in the form of underground water. At all times, the atmosphere is the outer envelope, but it, too, penetrates and influences the other two. The three interwoven make a fourth sphere possible, the biosphere. This is the thin coating of living organisms which penetrates and lives in all three. Our Earth is both dependent on these organisms and influences them. While Earth science is not concerned with living things, it is often impossible to explore topics like soil formation without recognizing the existence of this fourth sphere.

RELATED CLASSROOM ACTIVITIES

Earth science can contribute a great deal to the understanding of geography. Children who live near a river can gain a better idea of a watershed and the influence of the whole river by seeing it in miniature.

In practically every county in the United States where some land is used for crop production, teachers can obtain specific information on the soil types and problems, as well as water management and use of the area, by contacting their Soil Conservation Service, an agency of the U.S. Department of Agriculture. The Soil Conservation Service is active in towns, suburbs, and cities, and is playing a new role in regional planning by providing advice on the best part of the local lithosphere to use for the construction of things like a sewage treatment plant, a municipal building, or a concrete play area.

TEACHER PREPARATION

While a teacher may make new discoveries by walking around the schoolyard in the rain, a lot of preplanning for studying water can be done on a clear day by observing the marks from the action of water and then deciding how to set up activities when it does rain.

FIELD TRIP POSSIBILITIES

1. During a rain, collect rain water by putting a funnel in a jar in an open area on the grounds; at the same time, collect water that drips off different

places like the roof of a building, a fire escape, a stone wall, and a tree. Immediately following the rain, use a meat baster to collect water which is standing or running in places like a clay play area, a street gutter, or a hollow place on a pavement or rock. Take the samples indoors and put equal quantities of each in clean jars labeled with the water source.

Examine them for color, odor, and turbidity. List each sample on a chart and enter your observations. Put the jars on a warm windowsill for several hours (warm water cannot hold as much air as cold water). Examine the sides of the jars for air bubbles. Add this information to your chart. Do all jars have the same number of bubbles? Where did the air come from? The following day, reexamine the jars. Has any material settled out of the water? Is any floating on top? How do the samples differ? Record your findings. Put each sample through a funnel with filter paper. Label each paper with the source of water and attach it to the chart. Put each sample of filtered water into a clean pan or beaker and boil it until the water evaporates. Examine the inside of the pan. The coating is dissolved minerals. Add your discoveries to the chart. What conclusions can the children draw?

2. During or immediately following a rain, take a field trip to observe waterways. A clay play area is particularly good for this; however, gutters of city streets sometimes illustrate river systems equally well. Let children discover tributary streams flowing together to make a river. Observe the difference in the load carried by rapidly flowing water compared to slowly moving water. Can children find waterfalls, rapids, deltas, canyons, or outwash plains? What happens if they build a dam? Does the lake formed by the dam fill up with silt? If so, examine the watershed upstream. Is it protected by plants or is it exposed soil?

3. Every school site will have evidence of soil formation. Take a trip to inventory the variety of soil-making processes. Look for rocks or concrete dissolving under a drip from a roof or a rainspout; rocks breaking up in stone walls, buildings, steps, or pavements due to freezing and thawing, heat and cold, roots of plants, or other physical forces; leaves piled up and broken by wind, running water, or children's feet; leaves, dead wood, and other organic material decaying.

4. Running water tends to level the earth. It removes material from high spots and deposits it in low spots. In doing this, water wears down hills and mountains and fills ponds, lakes, and even the ocean. Plants fight this process by slowing water so that its cutting and carrying power is lost. Plants make the earth more absorbent so the water sinks in and holds the soil instead of running off. During a rain, take a field trip to observe the action of plants on water. (Make sure all children are suitably dressed for the weather.) After a rainstorm, compare the erosion in two areas which are alike in all respects except for the presence and absence of plants.

5. After a rain that follows a period of dry weather, go outside with a trowel. Dig holes in various areas to see how deeply the water has penetrated. See how many influencing factors the children can discover, for example: soil composition, the slope of the land, plants, animal activities, and organic materials. (Be sure to fill the holes before you return to the classroom.)

6. The atmosphere also shapes the lithosphere and helps in soil production. Examine the school grounds for evidence of rock material carried by the wind. Was some of the organic material—for instance, dead leaves, seeds, spores of moss, algae, or lichens—brought to the area by wind? Was some organic material fragmented by being blown against hard objects? If your area is sandy, is there any evidence of sand carried by wind scouring rocks or scratching windows?

7. The top layer of soil is generally rich in organic matter and is called topsoil. Lower layers are made of rock fragments and are known as subsoil. If there is an embankment at one side of your school ground, make a clean cut on its side. How deep is the topsoil? What does it look like? What does the subsoil look like? How do the plant roots grow in the topsoil? Are there roots in the subsoil? What do plants get from each area? Can children check relative moisture in each area? Take a series of trips to see if the appearance and texture of the soil is the same after a rain, a drought, and a windstorm.

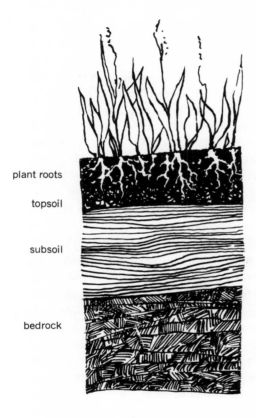

plant roots

topsoil

subsoil

bedrock

Soil Profile

The relative thickness of the topsoil, subsoil, and bedrock in a soil profile will vary, depending on the composition of the parent rock, the weather and climate of a region, the slope of the land, and the vegetation.

If there is no embankment, a hole 50 or more centimeters deep may be dug and the same soil differences may be observed. This exposed soil section is called a soil profile. It may be permanently recorded by cutting a piece of cardboard 20 centimeters wide and as long as the cut or hole is deep and coating a 8-centimeter-wide strip with tile cement. If this is pushed firmly against the profile and allowed to set for 24 hours, it can be peeled off and will have a coating of the soil materials. Space on either side of the mounted profile can be used for writing labels.

8. Earthworms come to the surface during or after a rain because water has filled their burrows and displaced the air. To see this displacement, remove both ends from two tin cans. Push one can down into the ground around an earthworm hole or holes; push the other into hard packed earth without earthworm holes. Put equal amounts of water in each can. Watch air bubbles come to the surface as the water sinks into the ground. Compare the number and size of bubbles in the two sites.

Geology

Rocks and minerals hold a fascination for both children and adults. For many people, there is excitement at discovering a fossil or a crystal, pleasure in the feel of an ocean-worn pebble, and wonder at a massive natural rock formation like El Capitan in California's Yosemite National Park, Great Stone Face in the Franconia Mountains of New Hampshire, or Natural Bridges National Monument in southeastern Utah. For others, the study of rocks, minerals, and rock formations opens the door to the long-ago past, to the pre-history, often the pre-human-story, of the Earth. Geology may be a highly technical scientific study, but it can also simply be a learning of general concepts which make all land forms more meaningful and exciting, as well as a bringing of new understandings to the geological processes currently going on.

Rocks, like other materials on Earth, are constantly changing. We classify rocks by the most recent story of their formation as igneous, sedimentary, or metamorphic.

Igneous rocks come from hot molten material. They may occasionally form from lava flowing from volcanoes and hardening on the Earth's surface, but much more commonly they come from molten material called magma, which is moving, cooling, and then solidifying far below the Earth's surface. It is estimated that 90 percent of the lithosphere is composed of igneous rocks. Most of this is the bedrock of the world: the solid continuous rock beneath oceans, the rock that forms the core of mountains, and the rock that is found beneath the thin coating of soil and other unconsolidated materials which geologists call mantle. Some igneous rocks are on the Earth's surface either as mantle or as rock outcroppings where bedrock has been exposed by erosional forces that have removed all the soil.

All rocks formed by the deposition of small particles and/or by chemicals in solution are called sedimentary rocks. When natural forces work on igneous rocks, they often become sedimentary rocks. For example, sedimentary rocks are formed when water gets in the spaces between the minerals (chiefly quartz and feldspar) which form granite, the most common igneous rock. As the water freezes and thaws, the rock disintegrates. Big pieces of quartz break loose, and break into smaller pieces of quartz; smaller pieces break into still smaller pieces which we call sand. At the same time, water and organic acids decompose the feldspars in the granite and form clays as well as compounds of lime, magnesium, potassium, aluminum, and sodium; oxygen also combines with the iron compounds, hornblendes, and biotite mica in the granite to form limonite (iron oxide) and other compounds. All fragments, new compounds, and new solutions are pulled downward by gravity and eventually carried, sorted, tumbled, and finally deposited by water. When these deposits are made in oceans, other forces may act to consolidate them and form new rocks.

Fine clay particles compressed by the weight of many layers of deposited material and dried out by the change in water level become shale, a layered rock which smells like clay when it is wet. Particles of calcium carbonate may be removed from the water by microscopic plants and animals which use the particles to make their shells. These minerals are eventually deposited on the ocean floor as the living organisms die and their bodies decay. These intricate microscopic shells may also be compressed by layers of deposits above them to form chalk. The English White Cliffs of Dover are the most famous example of this type of deposit.

Deposited particles are always cemented together by some mineral in solution. Shells of larger animals are sometimes cemented together by calcium carbonate in solution to form another type of limestone rock called coquina. Coquina is found along the Atlantic coast of the southeastern United States and in Puerto Rico. Much of the historic city of St. Augustine, Florida, is built of it. In still other situations, the calcium carbonate is deposited out of solution to form layers of limestone rock with no help from animals.

Deposited particles may also be cemented together by iron oxide, calcium carbonate, or less commonly, silica (quartz) in solution. Thus, both red and brown sandstone get their colors from iron oxides; sandstones which bubble when a drop of acid is put on them have been cemented by calcium carbonate. When pebbles are cemented together, we call the resulting rock a conglomerate. Concrete is really a manufactured conglomerate.

Metamorphic rock is the third type of rock. These rocks are formed deep in the Earth's crust when igneous and sedimentary rocks are chemically changed by heat and pressure.

At the same time that rocks on the Earth's surface are being broken down, moved, deposited, and reconsolidated, rocks under the Earth are also undergoing change. These changes are due to heat and pressure produced by the movement of hot magma and by internal forces of the type that produce upwarping and downwarping of the Earth. Generally, these movements are so slow that they cannot be observed. However, when they are rapid, they are accompanied by tremors, earthquakes, tidal waves, and other terrifying manifestations. It is not surprising that forces that inject molten-rock material between old rocks, forces that form mountains, fold and twist rocks, raise islands out of the sea, drop land into the sea, and otherwise dramatically transform the landscape should alter the rocks that are involved in the changes. These new rocks are, of course, the metamorphic rocks. If pressure is great but heat is limited, the metamorphic changes are physical. Shale sometimes becomes slate. Limestone sometimes becomes marble. These new rocks are more dense and durable than the rocks from which they were derived, but they are unchanged chemically.

Under other circumstances, when both heat and pressure are extreme, shale does not become slate, Instead, the minerals break down and recombine in new chemical compounds and a thin-layered metamorphic rock called schist is formed. Here the layering does not come from sedimentation, but rather from the arrangement of the newly formed crystals. Many schists are largely made up of mica. They may originally have been volcanic tuff, felsite, slate, or shale. Sometimes the layering formed in metamorphism is wide, from one centimeter up to many centimeters in diameter. The rock thus formed is called gneiss (pronounced *nice*). Gneiss is characterized by bands of feldspar, alternating with bands of darker minerals. Gneiss may be formed from sandstones, conglomerates, shales, slates, granites, or schists.

Some metamorphic changes occur over a large area and are called regional metamorphism. At other times, hot molten magma being forced between joints (cracks) in the bedrock changes the rock on either side. This is called contact metamorphism. The extent of these changes is dependent on many things: the temperature of the intrusion, the composition of the intrusion, the composition of the bedrock, the pressure, the speed of cooling, and the amount of gas and water present in the magma.

The changes may be as simple as the formation of a band of schist on either side of the intrusion or they may be highly complex, for example, when part of the bedrock melts and combines with the molten material of the intrusion and forms large crystals, metallic ores, or other significant structures.

In a mine, the intrusions of economically important minerals are known as veins. Contact metamorphism is often easy to identify on exposed rock structures. It can be seen on natural rock outcroppings, on the sides of quarries, on road cuts, or in deep excavations for buildings.

Locating, observing, and interpreting exposed bedrock can be a most exciting activity. Mountains are formed when rock masses slip along faults (fractures), when layers of rocks are upheaved (anticlines) or sink down (synclines), when rock masses are folded, even when valleys are cut into soft rock and leave resistant rock standing (such as the Catskill Mountains in New York State). Glaciers, water, and wind have all helped carve the landscape. The story of any city, town, park, farmland, or wasteland is partly a story of its geology.

RELATED CLASSROOM ACTIVITIES

For young children, rocks and minerals may be purely a sensory study. In nursery school, a pebble can be a beautiful thing to feel, to look at, and to play with. Getting to know one very distinctive mineral like mica in the classroom and then going outside and discovering it everywhere, in schist rock outcroppings and stone walls, in granite building trim, in concrete sidewalks and asphalt streets, is exciting to young children living in urban areas.

For older children, the more complex geological concepts can be fascinating. For all students, the study of geology should deal with examples from the world immediately around them, whenever possible.

Some areas have had books and pamphlets written about their geology. These can serve as good reference materials, but they are most valuable after the children have gone outside and made some discoveries for themselves.

The study of rocks and minerals or of Earth forces may be introduced in many ways: by a rock brought to the classroom, a geography lesson, a news item, a trip to an excavation, a new sidewalk, or the story of a national park.

One third-grade class in Manhattan that had watched alum, salt, and sugar crystals grow was very excited when they went outdoors and discovered garnet crystals in the veins of pegmatite in their local rock outcroppings. Even when they were examining the crystals, they were asking questions that would lead to further classroom discussion and research, such as, "How did they get there?" and "When?"

Garnet Crystal

Mica Crystal

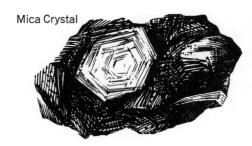

Garnet and mica crystals are widely distributed in metamorphic and igneous rocks. Both types of crystals range in size from the almost microscopic to many centimeters in diameter. Crystal size is related to rate of cooling; slow cooling produces large crystals.

TEACHER PREPARATION

Since all the Earth's surface is a product of geological forces and since geological forces are still at work, every school ground will have some evidence of the Earth's past as well as its future. Disintegration and decomposition can always be located. Sediments are being deposited on every school ground. Manufactured conglomerates, sandstones, and shales are everywhere. Igneous and metamorphic rocks like granite and marble are often parts of building trim and steps. Stone walls offer many teaching possibilities. Once a teacher goes outside and tunes in on the school site, he or she will find almost limitless possibilities.

For the teacher who wants additional information on local geology, contacting the geology department of the state university, the state department of parks, the local library, or the nearest natural history museum can be helpful.

FIELD TRIP POSSIBILITIES

1. Examine the rock outcroppings in your schoolyard. Note things like color, texture, crystals, hardness, the pattern in which the rock breaks and decomposes. What uses might these rocks have? If the class has been studying rocks and minerals, can they use their knowledge to identify the minerals? The rock?

2. Sometimes bedrock can be seen in the walls or floors of the basement. Enlist the custodian's assistance in planning a field trip "into the Earth." Observe the characteristics of these rocks. If there are also rock outcroppings on your school grounds, compare the rocks under the earth with those exposed. Are they the same kind of rocks? If so, do the outcroppings differ in terms of evidences of environmental influences? How many influences can children identify?

3. Take a field trip to discover all the ways in which rocks and minerals are used in your school building. Indoors, look for cut and polished rocks like marble or granite in walls, floor tiles, thresholds, steps, bathroom stalls, and behind fountains. And don't forget the window glass, a product of melted silica. Outdoors, look for things like stone curbings, building trim, and stone walls. Notice places where identifiable rock fragments are bonded to form a hard surface, like concrete walks, building blocks made of concrete, and black-topped play areas. Even the bonding agents, cement and asphalt, are geologic in origin. Cement comes from burned limestone and clay or shale. Asphalt is a liquid rock which occurs naturally in some places or is made from petroleum. Look for fused clay products like bricks, porcelain, and tiles. How many different kinds of metals can the children locate?

4. Many school areas have deposits of unconsolidated rock. Generally, these are a part of the geological story of the area. In many parts of the United States, they may have first been carried by a glacier, then tumbled along by the water from melting ice. Rocks like these have peculiar angles and flattened and rounded edges. Rocks that have been brought into an area by forces such as glaciers are called erratics. Have every child find a glacial erratic. Are they all alike? Group together the children who have rocks of the same color, size, and texture. How many different kinds of rocks has the class picked up? Are any like the local bedrock? Where did the ones that are different come from?

5. In an area where glacial or water-borne erratics are common, have the children collect as many different kinds of rocks as they can. Back in the classroom or sitting in a comfortable spot on the school grounds, divide the class into four teams. Let the children pool their rocks and arrange them in various ways: from smallest to largest, from lightest to darkest, by sharpness and smoothness, from softest to hardest, rocks with mica and rocks without mica, rocks with layers and rocks without, rocks that break easily and hard rocks. If the children have learned to identify rocks and minerals, let them group them by kind. Any rocks that cannot be identified could be set aside. As a special challenge, these might be identified by using rock and mineral keys.

6. Examine the rock fragments in sidewalks, driveways, and road surfaces. Are they a reflection of the local geology (for example, crushed rock from nearby quarries, pebbles deposited in the area by a glacier or by a local stream,

or crushed shells from a nearby shellfish industry)? Or were these rock fragments selected because of some physical property, such as: to provide roughness in icy areas or hilly areas with heavy rainfall, or, conversely, to provide a smooth surface for aesthetic reasons or to reduce friction? Are the same kind of fragments used in all the walks and roads around the school? If there are differences, is this a matter of economics, special planning, or is it unintentional?

7. To develop an appreciation for the formation of sedimentary rocks, examine a delta or outwash plain on your grounds (see *Earth Science* field trip 2, p. 116). Dig a hole in a delta. Can you see strata (layers) of different materials? Can you see why shale, sandstone, and limestone sometimes occur in alternate layers in areas of sedimentary rock?

8. Fossils are commonly found in sedimentary rocks. Examine mud puddles for footprints, worm trails, marks from raindrops, cracks from drying, and other things that are found as fossil prints. Can you find plant or animal materials trapped between the strata of the small deltas on your grounds? How would heat, pressure, or time alter them?

9. Dog, cat, pigeon, and sparrow footprints can often be found in the manufactured concrete we use for sidewalks. Compare them to photographs of dinosaur footprints found in the Connecticut River Valley in Massachusetts. What things can be learned from the sidewalk prints? What do we learn from the dinosaur prints?

10. When two different types of rocks occur in one area, one may break down faster than the other. This is called differential weathering. Thus, the sandstone is eaten away below the tough dolomite limestone at the edge of Niagara Falls in New York State; a resistant rock caps the mesas in the western and southwestern United States; a vein of crystals which formed in bedrock may extend above the rock surface or it may be sunk in the bedrock, depending on the relative hardness and softness of the two rocks. Examine the rocks in your area for evidence of differential weathering due to differences in hardness, solubility, or compactness.

11. Sometimes differential weathering is due to joints. These naturally occurring cracks in rocks make it easy for water, ice, and plant roots to act on the rock. If you have a rock outcropping with joints spaced at different intervals, compare the rock breakdown in places where the joints are widely spaced and closely spaced.

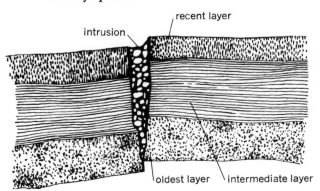

intrusion

recent layer

oldest layer   intermediate layer

Rock Layers with Intrusion

The large rock masses exposed by road building enable us to see the results of geological activities that occurred within the Earth millions of years ago. This diagram illustrates the shifting of rock layers that resulted from either the intrusion of hot molten material in a fault; or the intrusion of the hot material following a shifting of rocks along a fault.

12. If an area where bedrock has been cut through is available on the school ground or in the school basement, examine it for signs of Earth movements like folds, synclines, anticlines, and faults. Make models of your bedrock with Plasticine or papier-mache.

Earth Movements

This diagram illustrates some of the other geological activity which can be observed when bedrock is exposed by the construction of roads, quarries, or buildings.

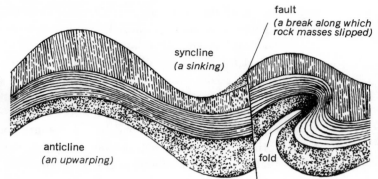

fault
(a break along which rock masses slipped)

syncline
(a sinking)

anticline
(an upwarping)

fold

13. Animal prints as well as fruits, seeds, and leaves that fall on snow all can contribute to fossil understanding.

14. Many geological processes can be observed in the snow. All the elements and many compounds on Earth can exist in three forms: gas, liquid, or solid. Water is the only compound that exists in these three forms at temperatures that occur naturally on the earth's surface. Others can be liquefied or turned to a gas or solid only by extreme changes in pressure and/or temperature. We define a mineral as a naturally occurring substance formed by inorganic (nonliving) processes, having definite physical and chemical properties and a molecular structure usually arranged in crystalline pattern. By this definition, snow is a mineral. As it drifts to Earth and forms layers, it is an unconsolidated sedimentary rock. Heat will melt it and, like magma, it flows, fills cracks, and when cool, hardens into an igneous rock. Heat and pressure on snow (as in making a snowball) or extreme pressure (as in a glacier) metamorphose it. The crystalline structure of snow, the layering of snow in successive storms, the change from sedimentary to igneous or metamorphic rock, can all be observed anywhere snow falls and remains on the ground.

15. To help understand the Earth's timetable and different time intervals, geologists use places where one layer of sedimentary rock was deposited on another. Sometimes the different layers are of different materials and are easy to recognize; however, they may be composed of the same materials. One way of recognizing different layers is by unconformities and disconformities. In an unconformity, the first sedimentary layers were folded, faulted, or twisted and pushed up in other patterns. Then a new layer of sediments was deposited on top of the contorted rocks. In disconformities, the top surface of the early rock layer was changed by exposure to disintegration and decomposition before another layer of sediment fell on top.

Both can be observed with snow. A snow plow tumbles blocks of snow along a road in different directions. A new snow storm puts a horizontal layer of snow over them—an unconformity. Snow melts, forms crust and icy spots. A new layer of snow covers them—a disconformity.

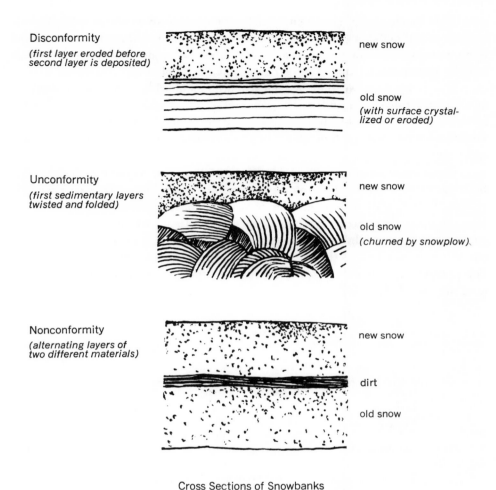

Disconformity
(first layer eroded before second layer is deposited)

new snow

old snow
(with surface crystallized or eroded)

Unconformity
(first sedimentary layers twisted and folded)

new snow

old snow
(churned by snowplow)

Nonconformity
(alternating layers of two different materials)

new snow

dirt

old snow

Cross Sections of Snowbanks

Examining cross sections of snowbanks can demonstrate how geologists use cross sections of rock to read the Earth's history.

These can often be observed better by cutting through the snow to form a cross section. A cross section may also reveal layers of different sediments such as fly ash, cinders, and other pollutants that were deposited between two snowfalls.

Weather and Climate~~~~~~~~

Although we have weather satellites circling the Earth, photographing and sampling the atmosphere with sophisticated instruments, and although one can dial an hourly weather report and prediction in all the major cities of the United States, conditions do not seem to have changed dramatically since Mark Twain issued his much-quoted statement that everyone talks about the weather but no one does anything about it. In fact, it sometimes seems as if the official prediction is no more accurate or not even as accurate as the prediction based on grand-father's rheumatism, Aunt Anna's bunion, or the local stock of weather lore.

This is because weather is a local matter; and while it stems from Earth movements, sun spots, and moving air masses, it is also partly influenced by factors like lakes, parks, forests, and acres of asphalt. When we talk about weather, we are talking about the condition of the atmosphere at a given place and time.

We describe weather with words like cold, hot, warm, cloudy, dry, humid, freezing, snowy, rainy, clear, stormy, or windy. Over the years, we have developed a variety of instruments for measuring the weather and for predicting the future weather. Some are complicated, but others, like thermometers, are part of our everyday life. Sometimes a simple instrument is almost as effective and much more useful to the layman than an expensive, complicated instrument. For instance, the Beaufort scale, reproduced here, is a useful chart for determining wind velocities.

BEAUFORT SCALE

Code Number	Description	Signs	Kilometers per hour
0	Calm	Calm; smoke rises vertically	less than 1
1	Light air	Smoke drifts; weathervane does not	1–5
2	Light breeze	Wind felt on face; leaves rustle; weathervane moves	6–11
3	Gentle breeze	Light flags blow; leaves and small twigs move constantly	12–19
4	Moderate breeze	Small branches move; papers blow; dust is raised	20–28
5	Fresh breeze	Small trees sway; crests form on inland waterway	29–38
6	Strong breeze	Telegraph wires whistle; large branches move; umbrella used with difficulty	39–49
7	High wind or moderate gale	Whole trees in motion; walking difficult	50–61
8	Fresh gale	Twigs break; progress of people impeded	62–74
9	Strong gale	Insecurely fastened parts of houses torn loose	75–88
10	Whole gale	Trees uprooted; buildings damaged	89–102
11	Storm	Widespread damage	103–117
12–17	Hurricane	Devastation occurs	above 117

When we talk about wind, we are talking about moving air. Air moves because of differences in temperature. As it moves, it also causes changes in temperature. In addition, wind picks up and carries water vapor, thereby influencing humidity. In other words, all weather is a series of interrelationships.

Weather in big cities often differs from the surrounding countryside for several reasons. Streets and buildings convert infrared rays into heat more effectively than grass, trees, and snow. Heat from steam pipes, automobiles, and buildings tends to raise city temperatures. Street canyons created by high buildings often serve as wind tunnels. Even the millions of people with their 37°C bodies influence the temperature and increase the water vapor content of the atmosphere with their breathing. However, this increase in humidity cannot offset the great loss of moisture created when plants are replaced with asphalt and concrete; therefore, cities are essentially arid places unless wind brings moisture in from adjacent lakes, oceans, or rivers.

Cities are not the only places where human activity has influenced the weather. The clearing of forests has caused water runoff with consequent negative effects on humidity and rainfall. In turn, this has caused an increase in temperature extremes due to the loss of the buffer effect of the slow heating and cooling of water and vegetation.

All of these changes are unplanned and indirect, but humans have consciously been trying to change the weather since the earliest times. About seventy-five thousand years ago, as the ice sheet moved over Europe, Neanderthal Man built fires in caves and thereby created small areas with their own weather or climate (microclimates) that were more comfortable than the cold, windy, tundra-like outdoors. Today we all live in microclimates created by walls, windows, furnaces, and air conditioners. But the weather outdoors is harder to change. Smudge pots may eliminate frost in orchards or ice on a landing strip for a short period of time. However, if the cold weather continues, the air becomes too chilled for artificial heat to be effective; then smudge only adds to the pollution, and fruits, blossoms, and runways freeze.

Through the ages, people have tried to induce rain in dry seasons by dances, magical incantation, and religious ceremonies. In more recent times, great hope was held for making rain by expelling tiny pellets of frozen carbon dioxide or particles of silver iodide in the clouds. Newspaper headlines in 1946 announced this scientific breakthrough as the eventual answer to the world's problems of drought. But rain-making can only work when there are clouds to be seeded. The little droplets of water must be there, ready to be nudged downward. Sometimes the clouds look right, the money is put forth, everyone is optimistic, the plane takes off, the chemicals are dropped, but nothing happens—or even more discouraging, the fluffy, hope-holding clouds disappear. Unfortunately, manipulated weather is no more predictable than naturally occurring weather.

Owners of ski slopes advertise that they can make their own weather with snow-making machines. Actually snow-making consists of spraying tiny droplets of water into the air when the temperature is below freezing. As the water particles fall to the ground, they freeze into tiny particles of ice. The same result could be accomplished by making ice indoors in a giant freezer and carrying it out on the slopes. But it is cheaper and easier to pump the water up the slope and take advantage of the existing weather conditions.

While snow-making takes advantage of existing weather conditions rather than changing them, the addition of many tons of ice to mountain slopes undoubtedly influences the weather in a variety of ways. Air chilled by the manufactured snow moves down to the valley as wind. The snow has a different effect on the Sun's rays than the bare Earth would have had. Melting artificial snow may change the humidity, influence precipitation, and serve as a source of ground water.

As long as the water used for making snow returns to the valley as precipitation or ground water in the warm months, and as long as the temperature changes are not harmful to the community, the use of snow-making machines will probably go unchallenged. This is not true of all twentieth-century activities. Some scientists today are deeply concerned because they feel the increase of carbon dioxide in the air is changing the world's climate.

Climate is a composite of weather conditions over an extended period of time. If the average temperature of the atmosphere dropped only six to eight degrees Celcius, snow and ice would begin to accumulate. Continental glaciers would again form, and a new world climate would develop. If the temperature rose only slightly, the water locked in polar ice caps and mountain glaciers would melt. The oceans would rise approximately 50 meters and many of the world's major cities would disappear under water.

By studying sedimentary rocks and fossils, we know that dramatic changes like this have occurred in the world's climate from time to time. We believe that one factor in these changes was an increase in carbon dioxide due to great volcanic activity. Until the last couple of decades, we believed that the atmosphere could not be changed by human activities because the atmosphere was so extensive. Today we are discovering that this is not true. Without meaning to, we may change our weather as we pollute our atmosphere.

RELATED CLASSROOM ACTIVITIES

It is hard to think of any subject which cannot involve some study of weather or contribute to the study of weather. We read about weather in historical events like England's battle with the Spanish Armada, Napoleon's defeat in Russia, and Washington's crossing of the Delaware. Weather and climate are also a part of geography. They influence the crops people raise, the food they eat, the way they dress, the places they live, and the houses they build. Heat, conduction, convection, the behavior of solids, liquids, and gases, and many other topics often studied in science classes are all aspects of weather. Newspapers and magazines report on unusual weather occurrences as current events. Making a six-sided snowflake can be an art experience. Dressing properly for weather is a matter of health.

TEACHER PREPARATION

Preparation for a weather field trip to the school grounds may only involve selecting the best teaching area. If the class is studying a specific phenomenon, it may be a matter of waiting for that kind of a day, or making sure that everyone knows the kind of clothing and footgear to bring for a trip in the rain or snow.

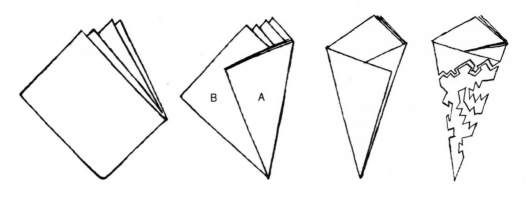

Paper Snowflake

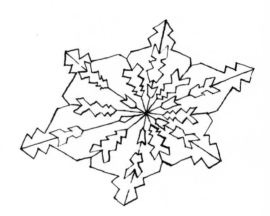

Snow crystals are six-sided. Models can be made by using small, white paper napkins that are folded in quarters. If using square paper, fold it in quarters. Hold the folded paper in your left hand so the four cut sides are on the right and top. Fold the section on the right to the center so that it forms a triangle (A) exactly equal to the un-covered area (B). Then fold B over A. Beginning about half to three-quarters of the way from the bottom point, cut a jagged line all the way across, completely cutting off the entire top area. Make cutouts on the sides. Now open the paper and you will see your own snowflake.

FIELD TRIP POSSIBILITIES

1. Wind can be experienced with many senses. Let children close their eyes and feel the wind. Can they smell the wind? Hear the wind? What things can they see that tell them about the wind? If they wet one finger and hold it above their heads, what happens? Does what they feel correspond to what they see?

2. If a single specimen of a tree or bush with wind-dispersed seeds grows away from the building on your grounds, divide the class into four teams. Have each team search in one direction to see how many seeds can be found on the ground in their quarter and how far from the tree the seeds can be found. Back in the class-room, their findings could be recorded as a picture or a graph. The figures on a graph will probably raise a new question: Why does the wind blow more often and harder from one direction?

3. Set up a weather recording and forecasting station on your school ground. You can make a simple barometer from a bottle, a tight-fitting cork with one hole, two 4-centimeter pieces of glass tubing, rubber tubing, wax, and cord. Fill the bottle with colored water. Fasten the two pieces of glass tubing together with the rubber tubing. Insert one piece through the cork. Put the cork in the bottle securely. Seal with wax. Invert the bottle. Suck the water into the tube. Tie the tubing to the neck of the bottle so that it forms a U. Hang the bottle in an inverted position in a cradle knotted of cord.

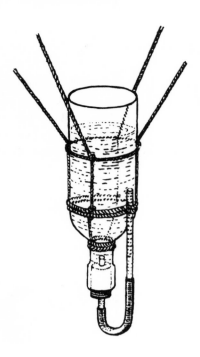

Barometer

When children make equipment like this water barometer and set up their own forecasting station on the school grounds, they gain an understanding of weather relationships.

When air pressure is high, it will push on the colored water in the glass tube and the water will rise in the bottle. When air pressure is low, the water level in the bottle will drop.

With this barometer, a thermometer, the Beaufort wind scale, and the school flag to tell wind direction, a daily weather report can be written. As data is collected, weather prediction may also be tried. If the class feels the need for more information, they can obtain a cloud chart from the United States Weather Bureau or build additional equipment as described in some reference books on weather.

4. Immediately following a hailstorm, hurry outside to find the largest hailstones. Try to cut or break a hailstone. Can you see the layered structure?

5. Microclimates can help with the understanding of weather. Compare the temperature and air currents in a grassed area and an asphalt area. What can this tell you about the weather in farmlands compared to the city? Can you find a microclimate that could be compared to a forest? A desert?

6. During a snowstorm, go out and catch snowflakes on dark colored paper, or, for longer examination, on black cloth dipped in water and frozen in the school freezer. Examine the flakes with a hand lens. The larger flakes were formed slowly in the lower, warmer clouds. The smaller flakes froze quickly in the high, super-chilled clouds. Can you tell where the snow is coming from? Notice that the larger flakes are more ornate, whereas the smaller flakes, which come from cirrus clouds, tend to be simple hexagons. (Sometimes flakes are wind-tossed and broken so their six-sided quality is destroyed.)

7. Snow and frost are frozen water vapor. Examine both. What have they in common? Ice and sleet are frozen liquid water. How are they alike?

Solar System ◟◞◟◞◟◞◟◞◟◞◟◞◟◞◟◞◟◞◟◞◟◞◟◞◟◞◟◞◟◞◟◞◟◞◟◞◟◞◟

Recently a group of teachers were talking about the knowledge and lack of knowledge which modern children bring to school. One third-grade teacher said, "Would you believe that in this space age I have children who think the Sun revolves around the Earth?" This may seem like an amazing situation until someone asks us personally to prove that the Sun does not. Both our speech and

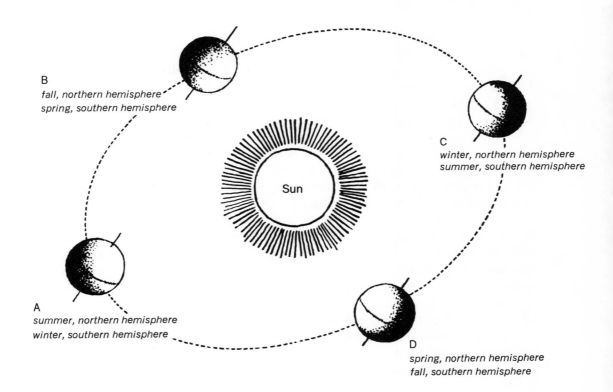

B
fall, northern hemisphere
spring, southern hemisphere

C
winter, northern hemisphere
summer, southern hemisphere

A
summer, northern hemisphere
winter, southern hemisphere

D
spring, northern hemisphere
fall, southern hemisphere

Earth's Rotation Around the Sun

The slant of the Earth on its axis results in seasons, due to the unequal distribution of the Sun's energy. Thus day at the north pole is 24 hours long at the same time that night at the south pole is 24 hours long. When the Earth is at positions B and D, every spot has a 12-hour day and a 12-hour night. If there were no slant to the axis, there would be no variation in day length and no seasonal change.

our daily observation would indicate that the Sun does move around the Earth. We say, "The Sun rises in the east and sets in the west." We watch it move across the sky.

It wasn't until the sixteenth century that Copernicus (1473–1543) advanced the theory that the Sun was the center of the solar system with the Earth moving around it. About one hundred years later (1610), Galileo discovered the moons of Jupiter with his telescope and became convinced from watching their movements that Copernicus was correct. A little later, he turned his telescope on Earth's moon and discovered that it did not shine with its own light.

This discovery was the real breakthrough, and even now explaining the phases of the moon is the easiest way that a layman with no training in astronomy or knowledge of celestial mathematics can prove that the earth moves around the Sun.

The moon shines with light reflected from the Sun. Since the moon is a sphere, one half is bathed in sunlight and the other half is in darkness. At one point each month, the side of the moon toward the Earth is directly across from the Sun so we see the whole lighted surface. This is the full moon.

At another point, 14¾ days later, the side of the moon toward the Earth is turned completely away from the Sun, and the moon cannot be seen at all. Astronomers call this the new moon, but new moon is really not a visible moon. Very shortly after this, however, as the moon continues on its orbit, a tiny edge of light becomes visible, and we see the crescent moon. As the part of the moon that we see gets bigger each day between new moon and full moon, it is described as a waxing moon. As it grows smaller between full moon and new moon, we call it a waning moon.

Eclipses of both the Sun and moon also help us understand the paths in which the planets travel. An eclipse of the moon occurs when the Earth's shadow falls directly on the full moon. An eclipse of the Sun occurs when the new moon is directly between the Earth and the Sun.

Because of the distances involved, the relatively small moon (diameter 3,475 kilometers) can blot out the Sun (diameter 1,390,000 kilometers).

If the moon traveled around the Earth on a regular orbit directly above the equator, we would have a solar or lunar eclipse each month. The moon's orbit, however, is elliptical, swinging high above and equally far below the equator with the result that eclipses at any given place are an irregular occurrence.

Today the moon's orbit has been so completely studied and plotted that eclipses can be predicted for centuries ahead of time.

To an imaginative and thoughtful child, the fact that the Earth is traveling about 53,400 kilometers per hour raises the same questions that people in the sixteenth and seventeenth centuries raised: "Why don't we feel it? Why don't we fall off?"

The answer to the first question is easily observed in many travel situations. You don't feel an airplane in motion on a smooth flight; you only know that it is moving by looking at the Earth "moving" below you. If you are sitting in one of two trains in a station and either starts to move, you may have to look at fixed objects out the windows on the other side to decide whether it is your train or the adjacent one that is in motion.

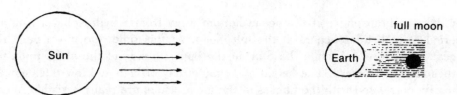

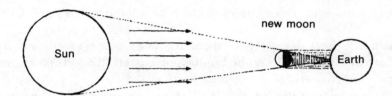

Eclipse of the Moon

Since the moon shines only from light reflected from the Sun, it is blotted out by the Earth whenever the Earth is directly between the moon and the Sun. To ancient peoples who knew nothing of the movement of the solar system, the Earth's shadow gradually creeping over the moon was a terrifying thing.

Eclipse of the Sun

When the moon gets directly between the Sun and the Earth, it blocks out our view of the Sun. The ability of the small moon to totally block out the large Sun can be appreciated by holding your hand at arm's length and using it to block out a distant object like a tree, a car, or a tall building.

When Earth moves through space, everything moves with it, so we have no feeling of moving. It is like being in an airplane. The plane moves at a set speed and, therefore, all the people and objects in the plane move at the same speed. As a passenger, you have no sense of motion. You only begin to feel motion either when the plane changes speed or when you look out the window. Since the Earth does not change speed and since we cannot see other planets passing with the naked eye, we do not feel the Earth moving.

We do not fly out into space because the Earth's pull of gravity is in perfect balance with the centrifugal force generated by the Earth's motion. Centrifugal force is the pull away from the center around which a body rotates.

By the same token, the Earth does not crash into the Sun because its gravity and centrifugal force are in balance with the Sun's gravity. In a like manner, the moon stays in orbit, neither crashing into the Earth nor rushing through space to the Sun because of the balance between the pull of gravity of the Sun, moon, and Earth and the centrifugal force generated by the moon as it travels around the Earth.

The pull of gravity of the Sun and moon can be recognized by the tides. The Earth's hydrosphere is more responsive to these forces than the dense lithosphere. The highest tides occur when the Sun and moon are both pulling in the same direction. This occurs at noon at the time of the new moon. On the opposite side of the Earth (midnight), there is also a high tide because the Sun and moon are pulling the lithosphere of the ocean bottom away from a hydrosphere. Another period of high tides occurs at the full moon. At this time, the moon pulls the ocean water upward while the Sun on the opposite side of the globe pulls the lithosphere downward. The height and time of day of high and low tides can be directly correlated with the phases of the moon which are really a visible record of its orbit.

Of course, our climate and our seasons are also basically the result of our peculiar relationship to the Sun. Our temperature range comes from our distance from the Sun and from the fact that the Earth rotates once every 24 hours. If the Earth rotated only once while we made one revolution around the Sun as both Mercury and Venus do, one side of the Earth would be unimaginably hot while the other side would be equally cold. (Think of the contrast in temperature on the two sides of the moon, which has days and nights the length of 14¾ Earth days.)

Mars has a 24½–hour day, so its temperature range must be moderate, but because of its greater distance from the Sun (approximately 228,500,000 kilometers), its climate is considerably cooler than Earth's.

Our seasons arise from the fact that the Earth slants on its axis. Therefore, the Sun's rays strike different parts of the Earth at different angles while it orbits around the Sun. Where the Sun's rays strike the Earth at a 90-degree angle (directly perpendicular to the Earth), the highest possible temperatures are found. This is due to two factors: First, there is a greater concentration of heat striking one spot, and second, a daily 12 hours of darkness does not permit the cooling that occurs at the poles where the Sun never shines for weeks.

RELATED CLASSROOM ACTIVITIES

Size, distance, and speed in the solar system are all things that are sometimes difficult for youngsters to comprehend. Visual classroom experiences, however, can help students understand these concepts.

Models of the solar system can be made in various materials. A flannel board showing the Earth, moon, and Sun can be used to illustrate relative positions and size as well as demonstrate how eclipses and tides occur. The flannel board can be available for children to return to repeatedly until they understand position relationships. Questions may be posted on the board such as: What is the tide like at the time of a total eclipse of the Sun? Why are we having a total eclipse here while the children of Peru are watching a partial one? (Note that you can make Earth—diameter almost 13,000 kilometers—and moon—diameter 3000 kilometers—to scale; but even if you make Earth equal one centimeter, the Sun's diameter of 1,390,000 kilometers will be 101 centimeters. If Earth is 0.5 centimeters in diameter, moon 0.125 centimeters, the Sun will be 50 centimeters; and they will fit on the flannel board. However, the children may decide to use a bigger scale and draw the Sun on a chalk board, or make a paper Sun and line it up with the flannel board. Just working out how to make the flannel board can be helpful in learning size concepts.)

With the current interest in space explorations, flannel board models of various planets could be made. For Mars, there could be cutouts showing the polar caps, the bright areas (dry, desertlike regions, yellow-orange in color and covering about two-thirds of the planet's surface), the dark areas (irregular patterns, greenish or bluish grey in color and covering about one-third of the planet's surface); for Venus, a representation of the dense atmosphere; for Jupiter, an indication of its 16 moons. As discoveries are made, other features could be added.

Three-dimensional models are even more dramatic for appreciating relative size if the whole solar system is being studied. The small planets may be made of substances like papier-mache, Plasticine, or clay, but larger planets will need to be built on chicken wire, crumpled newspaper, a small box, or some other armature because, again, with one centimeter equaling 12,800 kilometers, Jupiter will have an 11-centimeter diameter. If styrofoam blocks are available, scale models may be made from them. The Sun, with its 108-centimeter diameter (on a scale of one centimeter equals 12,800 kilometers), cannot be built in three dimensions in a classroom, but it can be cut out of cardboard and mounted on a wall, or it can be built outdoors.

If the class uses the same scale to figure distances from the Sun, the vastness of space will be dimly sensed. Thus, the Earth's distance of 150,000,000 kilometers at one centimeter equal to 12,800 kilometers puts the earth nearly 300 meters away from the sun. On this same scale, Pluto would be 12.5 kilometers away. This distance might best be understood using local landmarks. In relation to the school, Pluto might be at the post office, Venus at the supermarket, and Mars at the ice-skating rink. When illustrated in this graphic manner, it is easy to understand why Pluto could not even be seen with a telescope, but was discovered on sky photographs when astronomers began searching for another planet that they felt must be there because of the way in which Neptune's orbit was deflected.

Traveling at 300,000 kilometers per second, it takes the light from the Sun between five and six hours to travel the 6,000,000,000 kilometers to Pluto and practically the same length of time to be reflected from that planet to Earth. Compared to the distance from the Sun to Pluto, the Earth's distance to the Sun is very short, only 150,000,000 kilometers, and can be traveled by light in 8 minutes and 20 seconds. The great distance and the dim light reflected from the small (approximately 6,500 kilometers in diameter) celestial body we call Pluto made its study difficult even with the best of telescopes and photographic equipment. As a result, its whole orbit had never been plotted, and astronomers were not sure whether it was truly a planet, a moon of Neptune, or perhaps a once-upon-a-time moon of Neptune that got pulled away when centrifugal force did not equal gravity and is now a planet. Only after Voyager II passed Pluto, taking photographs and measurements, were they sure Pluto was a planet.

Perhaps the most helpful of all models are models that explain the seasons. All that is needed is a light source and sphere through which a stick can be inserted to make an axis. The light source may be an electric bulb or a flashlight with bulb exposed on the sides. Draw a circle around the light source to represent the Earth's orbit. The Earth may be a styrofoam ball, an old tennis ball, an apple, or any other easily available sphere. Insert the stick. Determine north and slant the stick (representing the axis) toward the north. Always keep the axis slanted in this direction. Rotate the Earth in various spots along its orbit. Find the place where the north pole is in the light for the complete rotation. This is summer in the northern hemisphere. Note the lack of light at the south pole; the southern hemisphere is having winter.

Locate the first day of spring, the first day of fall. Notice that both north and south poles have twelve hours of light on these days. Mark the first days of summer, fall, winter, and spring on the orbit. How are they spaced?

TEACHER PREPARATION

Preparing for school ground field trips in astronomy generally consists only of selecting the best teaching spot in terms of space and noise or, where shadows are involved, selecting an area of bright sunlight or one that combines sun and shadow. For daytime moon study, an almanac can be used to check the time the moon rises and sets.

FIELD TRIP POSSIBILITIES

1. The question, "Why don't we fall off?" can best be answered by taking a half-filled pail of water outside. If the pail is held at arm's length from the body at about waist-level and rapidly whirled in an upward circle, the water will stay in the pail (centrifugal force). If the whirling is slowed or stopped, when the bucket is upside down, gravity takes over.

2. One of the best ways to learn about the Sun is to observe shadows. For pre-primary and primary grade children who have never explored shadows, this can be a simple discovery of shadows, the things they do, and their cause. Put children in a circle and let them count the shadows in the circle. Does everyone have a shadow? Let them all find their shadow. What else has a shadow? Which way do all the shadows point? A shadow is really a no-sun-spot. Find the biggest no-sun-spot on your school ground. What made it? Find the smallest no-sun-spot. Can children make their own no-sun-spots bigger or smaller? Can children see that the amount of sunlight that an object blocks is related to the size of the object?

3. Older children can measure many large objects with shadows, since the size of a shadow is directly proportional to the size of the object. Therefore, if a 1-meter ruler held perpendicular to the ground throws an 66-centimeter shadow and a tree shadow is 5.5 meters long, the tree will be 8.33 meters tall $(66:100 = 5.5:x)$.

4. Shadow games help small children understand both shadows and the Sun. Put children in a circle and ask them to make their shadow touch their hand, dance, be very still, stop touching one foot, and stop touching both feet. The children can also try to make a shadow climb a wall, stand on a tree shadow, and make their own shadow disappear. Can children see that if the building has already blocked the sunlight there is no light for them to shut out with their bodies?

5. Go outside on a cloudy day. Look for shadows. Everything is under the big shadow made by the cloud.

6. Let children stand in the bright sunlight so their shadow is behind them and hold a thermometer in the light until the fluid stops moving. Note the temperature. Then have them turn in the opposite direction and hold the thermometer in the same position in their shadow. Wait for a change in temperature reading. Objects that block the sun shut off heat waves as well as light waves.

7. Using hands for thermometers, compare the heat in places where the sun is blocked with the same place in sunlight (for example, the two sides of a tree trunk, asphalt under the seesaw and beside it, grass in the shadow of the building and in the sun). Can children understand why nights are generally cooler than days at the same time of the year? What about the temperature on cloudy days, when clouds intercept the Sun's rays, as compared to bright days?

8. The Sun's position in the sky may be studied by observing shadows at intervals throughout the day. With chalk, outline the shadow of an object at 10 A.M. Return and outline the shadow of the same object at 2 P.M. This may raise questions that involve coming out every hour on the hour the next day. Shadows not only change direction as the relative position of Earth and Sun changes, they also change size as the angle of light reaching the Earth changes.

9. Make a sundial by marking a shadow of a fixed object like a flag pole or a stake driven into the ground every hour. This can be done by driving numbered pegs into the ground or by marking the end of the shadow with stones with the hours marked on them.

10. The primitive sundial discussed in trip 9 above will soon be inaccurate. (It is exactly correct only on the day when it is made.) For an accurate sundial, construct a gnomon (the pointer of a sundial) with an angle equal to the latitude of your school. Point it due north. Set it up and mark the hours.

11. The primitive sundial was inaccurate because the Sun's rays strike the Earth at a slightly different angle each day. To note this change, draw the shadow of some object at the same time once a week. (Make sure you choose an hour that remains daylight even if the area goes on or off daylight saving time.)

12. Make a study of the daytime moon. During what hours is it visible? What shape is it? Back in the classroom, use the flannel board to figure out the reason why a full moon is never visible during school hours. What other phases are not visible during school hours? Why? The faintness of the moon in the daytime is simply a matter of being drowned out by the bright sunlight. Using a small study lamp to represent the Sun and a mirror to represent the moon, reflect light onto the ceiling (the Earth). Without changing the Sun or moon, turn the overhead lights on. What happens to the patch of reflected light? (The full moon blots out the stars the same way.)

13. Go outside with field glasses or a telescope if one is available. Look at the moon; notice the craters and peaks.

14. Sometimes a crescent moon will have the faint outline of a full moon in it. The crescent is lighted by the Sun. The faint moon surface is lighted by Earth shine—sunlight reflected off the Earth onto the moon.

After this light story has been observed, children can work out the relative positions of Earth, Sun, and moon with charts, models, a flannel board, or with a light and mirrors.

Even a "temporary pond" like this mud puddle on the edge of a playing field can provide opportunities for studying forces like erosion, sedimentation, stratification of deltas, evaporation, the effect of rain on grassed and bare soil, and temperature differences including the rate of heating and cooling in different habitats over the course of a day.

Ecology

April 22, 1970 was a historic day. Not just because all kinds of people all over the United States joined in celebrating Earth Day, but because it focused public attention on the seriousness and complexity of our environmental problems. Ecology and environmental science suddenly became household words.

That does not mean that the concepts of Earth Day I were born in 1970, nor that our problems were solved then. Far from it. There have always been some persons sensitive to the interrelationship of the living and the nonliving world and concerned with the wise use of soil, water, and other resources. The number of concerned people has grown steadily in recent years as science and technology have greatly increased the potential for widespread destruction.

When Thomas Jefferson wrote and spoke about crop rotation and careful management of the soil in the late 1700s, few people listened because there were millions of acres of land available for the taking whenever a farm wore out. During the next 150 years, increasing numbers of people became concerned about land use, but it took the dust storms of the 1930s to usher in the Soil Conservation Service, an agency of the U.S. Department of Agriculture, and bring about an almost universal program of good farmland management.

The same story is characteristic of the management of every one of our resources. In colonial Pennsylvania, a law provided for planting trees to replace ones that were cut down. This, of course, is not adequate forestry management, but it was a reflection of the fact that as long ago as the seventeenth century, some people were aware of the need for growing trees as a crop. Needless to say, settlers who were surrounded by great forests looked on trees as expendable; to them, forests harbored wild animals and unfriendly Indians, and occupied land which was needed for crops. Later, forests became a source of wealth as wood was harvested and sold both here and abroad. And, as in the case of the land, the forests seemed limitless. Why worry about depletion when there were always more forests and land in the west? But 250 years of careless cutting, slashing, and burning brought us to the Pacific. It also brought us floods, erosion, land destruction and siltation of lakes and reservoirs as rain water ran unchecked over the mistreated lands. Finally, during the presidency of Theodore Roosevelt, the United States Forest Service was established with Gifford Pinchot at its head. Shortly after this, big lumber companies became aware of the fact that harvesting with no thought of the future was equivalent to killing the goose that laid the golden egg. Today, good forestry practice is recognized as enlightened self-interest.

Wildlife management has had a similar history. The fur trade represented one of the most important sources of income in colonial America and was one of the main incentives for exploration. Game seemed limitless. The people from the highly cultivated farms and cities of Europe had no sense of the interrelationships of soil, water, and wildlife, or of predator and prey, of habitats and reproductive potential. The Eskimo legend that said "The wolf and the caribou are one, for the caribou feeds the wolf and the wolf makes the caribou strong," would have had no meaning to these men. They saw wolves and other predators only as competitors and the destruction of any species in any given place as unimportant, since, they believed, there was always more to be had at the foot of the western rainbow.

It wasn't until the early twentieth century that men like William T. Hornaday, first director of the New York Zoological Society (and the author whose work helped save the bison from extinction), and others began to push ideas of wildlife conservation, and people began to band together in groups like the Audubon Society (founded in 1905) to enact legislation to protect birds and mammals.

Even with this increased awareness, we arrived at the mid-twentieth century talking about soil conservation, water conservation, forest conservation, wildlife conservation, mineral conservation, and conservation of scenic resources, as if each resource could be put in its own little packet and managed independently of the others.

But any farmer who had drawn up and worked under a conservation farm plan realized that when he was dealing with soil, he was also concerned with water, plants, wildlife, and even recreation. A farm pond, for instance, might provide water for irrigation, for spraying orchards, for livestock, fishing, swimming, and fire protection, while it checked runoff, reduced erosion, and raised the water table. This was the concept of multiple-use.

Undoubtedly, the best and biggest multiple-use project of the first half of the twentieth century was the series of dams built by the Tennessee Valley Authority (TVA) during President Franklin D. Roosevelt's administration. These dams were primarily built for flood control and to provide a steady water supply for navigation, but they were planned in such a way that they could also be used to produce clean electricity (made with water power instead of fossil fuel or nuclear energy).

In order for these dams to be effective, siltation had to be kept to a minimum. Agencies like the Soil Conservation Service in the Department of Agriculture and the United States Forest Service in the Department of the Interior worked together to clothe the mountains with trees and to introduce good farming and forestry techniques to keep soil on the farms and mountainsides. Educational programs were instituted to involve people and to help them reap the full benefits of the project.

While the concept of interrelatedness was developing on farm lands and other land-use projects, technology was making giant strides. Only half of the people in the United States can remember (and all of us find it hard to believe) that frozen foods and nylon were both inventions of the 1930s, that detergents, DDT, weed killers, and the first TV broadcast came in the 1940s, that aluminum foil, polyethylene, and other plastics entered homes in the 1950s. Mechanical developments followed a similar time chart from the Model T Ford of 1930 to the jet planes and moon explorations of the 1960s.

The western frontier was replaced by scientific discoveries. More and more people were saying, "Why worry? We can always synthesize it." It was popular to smile indulgently or joke about conservationists, bird watchers, and nature study.

In 1962, when Rachel Carson wrote *Silent Spring,* a provocative study of the dangers involved in the use of insecticides, many people began to question some of our insecticide programs for the first time. But just as many people tossed it off as one person's hysterical propaganda. And many of those who responded with concern gave a sigh of relief when chemical companies published slick ads showing how safe insecticides were when used correctly, thus enabling them to continue to eliminate all insects and other "vermin" with a free conscience.

But mass poisoning of the air and water cannot be ignored indefinitely. In 1950, conservation textbooks, if they mentioned them at all, listed the atmosphere and the oceans as natural resources which we really did not need to be concerned about in their totality. They stated that while local damage admittedly existed along shore lines and over industrial complexes, the diluting factor of the atmosphere and ocean was so great that polluting them on a grand scale would be impossible. But when DDT turned up in the bones of penguins in the Antarctic, carried by ocean currents from the continents of the northern hemisphere, when people died of smog in England, Japan, and the United States, and when truck farmers lost crops to air pollution, it became evident that we live in a total environment where all things interact and are interrelated. What happens any place affects us all. People were finally ready to listen to the ideas expressed on Earth Day.

But listening is not enough. We must take responsibility.

We must take much more responsibility than has been taken in the 20-year span between Earth Day I and Earth Day II on April 24, 1990.

In fact our record during that period is abysmal. Even though we have banned the sale of DDT and a number of other persistent pesticides in the United States, we still permit these chemicals to be produced here and sold worldwide. Too often Congress has yielded to the demands of industry and economics rather than the ecology. Our government has failed to realize that persistent pollutants are not restricted by boundary lines on maps; they travel on air and water currents all over the globe.

In the last 20 years, we have spent unimaginable sums of money on space exploration and have seen wonderful pictures of our beautiful blue water planet from space along with dramatic, mind-boggling pictures of the rest of the solar system. If there was any doubt before, one thing has become certain: Earth is unique. Only on this small planet can life as we know it exist and for that matter any observable kind of life. We must work to preserve this unique world.

We have many lessons to learn if we are going to protect our world. We must take some of these lessons from the Earth itself. Earth is the great recycler. The water cycles, the oxygen/carbon-dioxide cycle, the nitrogen cycles, the disintegration and decomposition of organic and inorganic substances, the variety of living organisms, producers, consumers, decomposers and their interdependency all work to maintain a healthy Earth. We must learn to recycle our waste as the Earth has.

Many areas now have recycling programs. Operated by the states, municipalities, and sometimes private industry, these programs offer unique learning opportunities. Our children must learn to recycle through participation. When one eastern city recently exempted schools from participation in a rubbish-sorting requirement, people protested the exemption because of the bad example it taught the students. At last, we are beginning to realize the importance of teaching by example.

Recycling is only one of the four R's of solid waste reduction. The others: Refuse, Reduce, and Reuse are even more important. These are measures that each of us can take to help reduce the problem of solid waste. Caring for the Earth must start in our homes and schools. Do faucets run unchecked? Do we waste paper? Does all our solid waste go to the garbage dump? Our children will learn more about the importance of our ecology through our example than any other way.

We can also teach our children to recycle by teaching about the Earth's wonderful recycling system. The natural processes of the food chain—decomposition and disintegration—are a part of our everyday world. Many of these aspects can be studied on the school ground. As children come to appreciate the wonder and beauty of Earth's natural processes, they can also learn more about caring for the Earth. Even the smallest school ground can open doors to world-wide relationships. Where does oxygen come from if we live in the city where people far outnumber oxygen-producing plants? What happens to the oxygen in areas where plant activities stop for several months because of winter or a dry season? A wind study can give us new appreciations of the importance of tropical rain forests and of a healthy ocean.

For children, the discovery of the role of wind in delivering the air we breathe can be exciting. It fascinates young children that the air they breathe can come from faraway. Teenagers may be interested in the destruction of the rain forest as a political and economic problem. They may even look to their own contribution to the problem as consumers of cheap rain forest beef at quick food stands. Wind studies and the relationship to the rain forests can lead to further studies on our use and misuse of trees. In an effort to preserve the forests, one third grade class challenged a whole school to conserve paper. They encouraged using erasers instead of throwing paper out, using both sides of the paper, and collecting papers for a recycling program. They also reduced consumption of paper towels and cups with posters on the dispensers that said "Help Save Our Trees: If one will do why use two?"

Projects such as these are wonderful for children. Children need to know that we can bring about change. Too often today they are bombarded with news of environmental crises and problems. These projects not only offer hope and give children a sense that they are contributing to the world, they also encourage understanding and respect for Earth's ecology.

Often the classroom, the school, and the school ground provide opportunities for students to put their ecological knowledge to work to become involved in bringing about change.

There is no better way to give young people positive feelings about themselves and the future than to foster this kind of activity.

TEACHER PREPARATION

Preparing for ecological thinking and teaching on the school ground is a gradual process which can best be accomplished by repeated trips. Teachers who have used the school ground for field trips all through the year will be amazed at how many interrelationships they have come to discover and appreciate.

FIELD TRIP POSSIBILITIES

1. Locate the part of your school ground which has the largest population of living things (other than humans). Make a study to determine all the factors that make this a better habitat than other areas. Look for nonliving factors like soil quality, moisture, sunlight, protection from wind, and pollution; look for webs of living things like insects that live on plants, and animals that live on insects; look for human factors like cultivation, fertilization, or protection from running feet.

2. Find a place where plants are dying. Try to determine the cause. Is this related to human activities; for example, salting walks in winter, cutting initials in bark or taking short cuts; Earth science conditions, such as too little soil in a crack in a rock, or too little moisture; animal activity, like too many insects, or rodent bark damage; other plants causing shading, disease, or crowding; or a combination of these factors?

3. Take a field trip to determine ways in which you might improve the environment around your school to make it more pleasant for plants, people, and other animals. Take steps to make some of these improvements. Do you sometimes have to make choices? Does the definition of conservation, "the wise use of resources for the largest number for the longest time," help in making these choices?

4. Look at your school ground as a piece of land to be managed for multiple use. Could it be improved by making any additions or changes? Would there be less sound pollution if the playing field were in some other area? Would new plants provide an outdoor laboratory, add beauty, improve the temperature, or reduce the pollution? Could plants be used to hold soil and check gullying? Should an area be set aside for gardening?

5. Photograph or draw scenes on your school ground to illustrate John Storer's *The Web of Life* (New American Library) or some other book on ecology. Write captions and then make an ecology exhibit.

6. A teacher in Springfield, Massachusetts, introduced the idea of ecology by discussing the words *relation* and *relationship,* then sending her class out to make a list of all the interesting things they could find on the school ground. When they returned to the classroom, the names of the objects they had observed were written on the blackboard, then lines were drawn connecting things that the children felt were related. As the discussion progressed, relationships grew into interrelationships. The pattern resembled a web—the web of life—ecology.

7. On streets with heavy traffic, a concentration of dirt can sometimes be seen on the lower stories of buildings, visible evidence of pollution from motor vehicles. One of the most destructive gases, carbon monoxide, is colorless and odorless. It damages plants (notice trees that grow where buses idle their motors). It also gives people headaches and, of course, in quantity, kills.

Take a field trip to discover ways in which automobiles cause pollution. Check the hood of a car that has just parked for thermal pollution. Consider the effect of thousands of automobiles on the temperature of a city. Compare the temperature of black-topped streets to that of grass. Notice spots on streets where oil, gas, and other substances have leaked. Shoes, rugs, and floors all suffer and deteriorate because of these products. Notice clouds of smoke and unpleasant odors from some exhausts.

8. Urban areas with a high concentration of dogs often have a greater manure problem than rural areas. Dog manure, like all manure, is unaesthetic; it damages rugs and materials, may spread disease and worms to other dogs or even to people, and does serious damage to plants.

Are dogs a problem on your school ground? Do people curb their dogs? Has anything been done to street trees to protect them from dog manure? Is it successful? If your school ground has a dog problem, can anything be done to improve conditions?

A few places have special areas set aside as dog toilets. These are circles of concrete which slant toward a central opening to the sewer. Running water is sprayed over the area at regular intervals, carrying the feces into the sewer.

9. Fly ash and other solid air pollutants can be caught on pieces of glass coated with petroleum jelly or another greasy substance. Studies of the varying amounts of pollution can be made by placing fresh plates for a given period of time in different areas of the school ground, at different levels of air, in exposed areas, and areas under trees and bushes; or studies can be made of pollution levels at different hours of the day or days of the week. If there is a marked difference at one time or place, can children find the cause? Can anything be done about it?

10. Snow is an excellent substance for showing pollution. When snow falls, the world is temporarily clean and beautiful. In places of low population and little pollution it remains that way; but in other places, our bad environmental housekeeping becomes very obvious. A field trip may be taken several days after a snowstorm to observe pollution. Look for places that are exceptionally dirty or clean. Look for sources of pollution.

The buildup of pollutants may be observed (and preserved) by collecting and melting one cup of surface snow from a given area each day as long as a snowfall lasts.

If the water samples are put in numbered jars and set on the windowsill, a gradual darkening will be obvious. Other things may also be observed and discussed, such as the type of pollutant, substances that float, sink, or dissolve, wind direction, and sources of pollution.

11. Ragweed pollen is a natural air pollutant to which many people are sensitive. If there is ragweed on your school ground, learn to identify it and remove it in the spring before it has a chance to bloom, then plant marigolds, crown vetch, lespedeza, or some other hardy plants to crowd out the seedlings which will come up to replace the ones you pulled. In other words, change the habitat to eliminate the problem.

12. Plants and animals that are not native to an area are called exotics. Often when plants and animals are introduced in a new area they become serious pests because they have no natural enemies to control them. Today, there are many laws prohibiting the bringing of exotics into the United States or even into a state or region without permission of the United States Department of Agriculture. For instance, it is illegal to sell or keep gerbils in some states in the southwest because if some of these small rodents with a high reproductive poten-

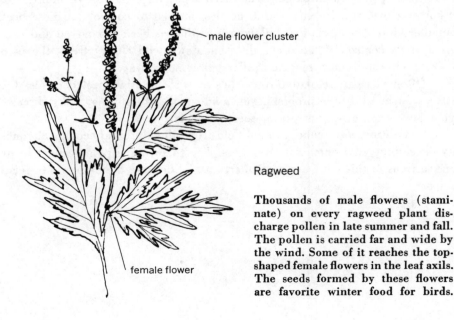

male flower cluster

female flower

Ragweed

Thousands of male flowers (staminate) on every ragweed plant discharge pollen in late summer and fall. The pollen is carried far and wide by the wind. Some of it reaches the top-shaped female flowers in the leaf axils. The seeds formed by these flowers are favorite winter food for birds.

tial escaped (as some pets always do), they would find themselves in a habitat much like their original home in terms of climate and food resources and might take over large areas because their natural enemies are absent. Some exotics that present problems in the United States are: house sparrows, starlings, pigeons, Norway rats, the giant African land snail that is causing much destruction in Florida, Japanese beetles, gypsy moths, San Jose scale, water hyacinth, fire ants, and carp.

Take a field trip to see if there are any plants or animals that compete unfairly with other living things on your school ground. Do library research to learn their history.

13. The web of life is a wonderfully interwoven pattern in which there are normal checks and balances. Whenever the web is broken, serious problems may arise. This is true of exotics which do not fit the new web they are in. It is often true when general pesticides are used which wipe out many things in the ecosystem. For instance, the effect of insecticides made of persistent hydrocarbons that kill all insects is also passed on to animals and may create more problems than it solves. Thus Gypsy moths may be killed, but leaf miners that eat the chlorophyll layers from inside a leaf, safely hidden from any spray, may cause as much damage as Gypsy moths when the parasitic wasps that normally keep them under control are also killed. When DDT was still being used in the United States, mouse populations soared to destructive numbers in some grain producing areas because DDT passed through the food chain and ultimately reached the rodent-controlling hawks. These useful birds were producing soft-shelled eggs and had died out in many places.

Sometimes we only learn about the delicate balance of the web of life when the fabric is torn.

Today many biologists are trying to understand and use the patterns of nature to improve the environment. Red ladybird beetles are introduced to control aphids; black ladybird beetles are used to control scale insects; milky disease, a bacteria that attacks only scarab beetles, is used to control Japanese beetles; another kind of bacteria that kills gypsy moths has been discovered and is being used in its control. When orchardists cut down wild cherry trees, the favorite food of tent caterpillars, tent caterpillars often move away.

Often you can see ladybird beetles at work as biological controls. If you have a Japanese beetle problem, you might look into getting milky disease for your lawn from a big nursery or seed company.

In addition, any school ground field trip will help build ecological concepts by developing environmental awareness. The following trips described in previous sections of this book are particularly useful because they stress interrelationships.

Other Ecology Field Trips

Topic	Field Trips
Plants	1,2,4–8,10, *pp. 17–19*
Trees	3,7,12,13, *p. 22*
Leaf Coloration	2–4,10, *pp. 26–27*
Seeds	8–10, *p. 35*
Grasses	3,5,6, *p. 38*
Dandelions	3,4, *p. 41*
Animals	2–4, *p. 45*
Vertebrate Animals	6,8, *p. 48*
Birds	2,4, *pp. 51–52*
Animal Tracks	3, *p. 55*
Insects and Other Arthropods	3,5,6,8,9,14,15,17, *pp. 61–64*
Earthworms	2–4,6,8,12,13, *p. 67*
Interdependence of Living Things	1–11, *pp. 72–74*
Characteristics of Living Things	8, *p. 78*
Requirements for Life	1–8, *p. 81*
Microhabitats	2–10, *pp. 83–86*
Population Explosion	1–9, *pp. 90–91*
Succession	1–3, *pp. 94–95*
Physical Science	
Physical and Chemical Change	6, *p. 103*
Heat and Light	1,2, *pp. 109–110*
Earth Science	3–5,7,8, *pp. 116–118*
Weather and Climate	2, *p. 129*

Cross-Referenced Listing of Field Trips
for Hard-Topped School Grounds

The field trips listed below can be carried out on school grounds without an inch of exposed soil.

Topic	Field Trips
Plants	1,2,8, *pp. 17,18*
Trees	14, *p. 23*
Seeds	2,3, *p. 34*
Animals	1,3, *p. 45*
Vertebrate Animals	1,10, *pp. 47,49*
Birds	1,2,4,5, *pp. 51,52*
Animal Tracks	1–7, *p. 55*
Insects and Other Arthropods	1,4,7,14,15, *pp. 61–63*
Interdependence of Living Things	
Characteristics of Living Things	1,2,7, *p. 77*
Requirements for Life	1,4,5, *p. 81*
Microhabitats	2,4, *pp. 83,84–85*
Population Explosion	7,9, *p. 91*
Succession	1, *p. 94*
Physical Science	1–8, *pp. 98–99*
Physical and Chemical Change	1–6, *pp. 102,103*
Sound	1–9, *pp. 105,106*
Heat and Light	1–10, *pp. 109–111*
Earth Science	1–3,6, *pp. 115–116,117*
Geology	1–6,9–14, *pp. 122–124*
Weather and Climate	1–3,6,7, *pp. 129,130–131*
Solar System	1–13, *pp. 136–137*

Most schools in an area with 75 or more centimeters of rainfall annually will probably have plants growing in cracks and crannies, even if they have a completely hard-topped school ground. If trees are growing either in areas broken in the blacktop or along the sidewalk, many other field trip possibilities exist, for there will be an opportunity for the observation of plants, which frequently grow in the exposed soil around the tree roots, as well as animals, which use the trees for food and shelter. Following is a list of these additional field trip possibilities.

Topic	Field Trips
Plants	3, *p. 17*
Trees	1–5,7–13, *pp. 21,22*
Leaf Coloration	1,2,4,10, *pp. 26,27*
Buds	1,2,4,5, *pp. 29,30*
Seeds	4–8, *pp. 34,35*

Topics	Field trips
Grasses	1,2,4,7, *pp. 37,38–39*
Dandelions	2,3, *p. 41*
Animals	
Vertebrate Animals	3,5,8, *p. 48*
Birds	3, *p. 51*
Insects and Other Arthropods	3,8,10–12,16, *pp. 61–63*
Earthworms	1–9, *p. 67*
Interdependence of Living Things	1,7, *pp. 72,73*
Characteristics of Living Things	3,4,6,8,9, *pp. 77,78*
Requirements for Life	2,6,7, *p. 81*
Microhabitats	6,8, *pp. 85,86*
Population Explosion	1–3,5,6, *pp. 90–91*
Physical Science	
Heat and Light	11, *p. 111*

Supplementary Materials

COURSES AND WORKSHOPS

The prospect of teaching science can be terrifying, especially for teachers with limited science backgrounds. Even teachers with a lot of science experience may be puzzled by an ever-changing world and the need to continue learning along with their students.

To fill these needs, in-service courses and workshops have been developed all over the country. Many local programs are now available, and information about them should be available through the principal's or superintendent's office. Other programs are national in scope, such as programs offered through the National Audubon Society and the National Wildlife Federation, or programs such as Project Learning Tree and Project Wild.

The National Audubon Society provides summer programs at various camp sites from Maine to the West Coast. While the two week sessions vary, all focus on learning in and about the outdoors. College credit is available for these sessions. Materials have also been developed for several courses of study: Living Lightly in the City (K–6), Living Lightly on the Planet (7–12), and A Place to Live (elementary grades with texts in English and Spanish). Special three-hour training sessions for using these materials are available in a number of U.S. cities. The National Audubon Society also supports Audubon Adventures, a school curriculum club whose materials are only available through club membership. Information on any of the programs above can be obtained by contacting the National Audubon Society, 950 Fifth Avenue, New York, NY 10022.

Likewise, the National Wildlife Federation offers several programs and services helpful to teachers. They offer a variety of outdoor study programs, from one-week summits to three-day Naturequests, located at a variety of environments in the United States. College credit is available for some programs. The National Wildlife Federation also publishes *Nature Scope*, a magazine addressing specific science topics from the perspective of teachers and adults who work with children, and *Ranger Rick*. For information on these publications and the nature workshops, write to the National Wildlife Federation, 1400 Sixteenth Street, NW, Washington DC 20036.

Project Learning Tree deals with land resources, stressing conservation education activities for K–12 in an interdisciplinary format. Teachers' manuals and materials for K–6 and 7–12 are available only through workshops sponsored by state departments of education. For information on schedules and localities, contact the department of education in your state.

Project Wild is a conservation and environmental education program emphasizing wildlife. The K–12 materials are only distributed through workshops which are conducted in states that have adopted the Project Wild curriculum. Until recently, Project Wild had a strong rural emphasis, but a growing awareness of the importance and impact of cities and the needs of urban children has brought about development of a broader-based curriculum. For information, contact Cheryl Charles, Director, Project Wild, P.O. Box 18060, Boulder, CO 80308.

Science Trade Books ~~~~~~~~~~~~~

Science trade books, like the school ground, are extremely important supplements to any science program. They can be used to introduce a topic or obtain supplementary information. By providing several books on the same topics, at different reading levels, a variety of supplementary information can be provided for children. Thus children can discover the joys of learning on their own as they glean information revealed to them as their developmental level permits.

With such large numbers of trade books published each year, choosing trade books for your school can be quite an undertaking. Two major sources of information on these books are the American Nature Study Society (ANSS) and the National Science Teachers Association (NSTA). ANSS is the oldest environmental education society in this country. Founded in 1908 by such educators as Anna Botsford Comstock, author of *The Handbook of Nature Study*, and Liberty Hyde Bailey, who was responsible for the establishment of the nature study program for children in New York, ANSS confers the Eva L. Gordon Award. Each year, they honor an author of children's science literature whose books reflect the accuracy, readability, timeliness, excitement, and interesting content that is so essential in science literature. Information on ANSS and the list of award winners may be obtained by sending a stamped self-addressed envelope to the American Nature Study Society, Dr. John Gustafson, 5881 Cold Brook Road, Homer, NY 13077.

Each year, the NSTA in cooperation with the Children's Book Council publishes an annotated list of 100 books appropriate for grades K–8. This list is broken down into eight categories: animals; biography; space and astronomy; anthropology and paleontology; life sciences, Earth science, conservation, and environment; medical and health; physics, technology, and engineering; and other.

The annual list is published in the March issue of *Science and Children*, a journal devoted to pre-school through middle school science teaching. Information on memberships and the programs of NSTA may be obtained by sending a self-addressed stamped envelope to the National Science Teachers Association, 1742 Connecticut Avenue, NW, Washington, DC 20009.

The John Burroughs Association also prepares an annual list of 20 nature books for young readers. While the books must be scientifically accurate to be included, the emphasis is on literature and the books are oriented toward language arts rather than on hands-on science. This list may serve as a lead-in or follow-up to science activities or just as an invitation to do more reading on the subject. To obtain this list, mail a self-addressed stamped envelope to the John Burroughs Association, 15 West 77th Street, New York, NY 10024.

Books for Teachers

The books discussed below are not presented as a syllabus or as a course of study, rather they deal with feelings and attitudes which can enrich your ecology program.

Keepers of the Earth, Native American Stories and Environmental Activities for Children by John Bruchac and Michael Caduto, Fulcrum Inc., 1988, 320 pp.

Almost every state includes some Native American activities in the social studies curriculum, but Native American studies can enrich and be enriched by many other subjects. In fact, it is a superb topic for an integrated curriculum. In this book, Abenaki Indian Joseph Bruchac has selected legends from 18 cultures while Michael Caduto has written supplementary materials. Caduto relates the basic concepts of the legends to today's world, provides basic scientific facts applicable to the legends, and finally develops a variety of child-tested hands-on activities to foster problem solving and responsible action. The book can bring together social studies, language arts, and science.

Listening to Nature: How to Deepen Your Awareness to Nature by Joseph Cornell, Dawn Publications, 1987, 95 pp.

This book, based on 31 quotes on nature from familiar authors, poets, and philosophers, is supplemented with activities and explanations designed to help the reader translate the quotes into personal experiences. Any reader will certainly emerge with a new awareness and a desire to share some of these sensitivity-expanding experiences with young people.

Creating Humane Climates Outdoors: A People Skills Primer by Clifford Knapp, ERIC Clearinghouse, 1988, 97 pp.

"Caring for people and treating the Earth gently both involve many of the same skills and attitudes" is the fundamental premise of this wonderful book. The text consists of a variety of structured activities to foster individual and group learning through direct contact with nature. The book is designed to help teachers and children get in touch with themselves, nature, and others through varied experiences that lead to shared experiences, creative thinking, and other intrapersonal and interpersonal skills.

Botany for All Ages by Jorie Hunken, Globe Pequot Press Inc., 1989, 157 pp.

The 126 activities in this book provide opportunities for hands-on discovery of basic plant facts and relationships. Pollution, insects, seasonal changes, flower parts, and self-pollination are examples of the variety of relationships, science concepts, and curriculum areas that can be selected for indoor and outdoor enrichment.

The City Kid's Field Guide by Ethan Heberman, Simon and Schuster, 1989, 48 pp.

Developed in association with WGBH Boston, producers of NOVA. Indoors, outdoors, vacant lots, backyards, parks, and downtown canyons—cities like every other part of Earth have plants and animals adapted to the environment. This book provides information on them as individuals and in terms of relationships. Charts, colored photos, and an index add to the usefulness of the book.

Early Childhood and Science by Margaret McIntyre, National Science Teachers Association, 1984, 136 pp.

For anyone dealing with little folk—2 to 6 year olds—this book is a gold mine of information on child behavior and of activities designed to capitalize and foster the natural scientific curiosity of children as they develop an awareness of themselves, of other living things, and of the environment.

Index